Big Qual

"This volume will be a milestone in qualitative analysis, but more than that it clearly demonstrates that big data can be qualitative as much as quantitative. The volume takes a unique approach to analysis – that of a depth and breadth archaeology. This is more than a metaphor and is the basis of a rigorous method of analysis that allows the merging of data, from multiple sources, or over time, to create emergent outcomes. This shows the enormous potential of 'qualitative big data' to move beyond traditional qualitative analysis to produce new types of data, that can challenge the simple quantitative-qualitative dichotomy. The approach shows the potential for richer understandings at the micro level, but also the possibility of wider social generalisation. The authors bring both a methodological sophistication to this work, but also their deep practical knowledge and experience. The volume can be approached, or used, in different ways – as a methodological treatise grounded in empirical research, or the individual chapters can be used as discrete modules in specific aspects of analysis".

—Malcolm Williams, *Cardiff University, UK*

"In *Big Qual: A Guide to Breadth-and-Depth Analysis*, Susie Weller, Emma Davidson, Rosalind Edwards, and Lynn Jamieson introduce a novel approach for working qualitatively with big datasets. Notably, big data has often been assumed to be quantitative in scope. Yet, in this book, Weller and colleagues aptly highlight the potentiality of qualitative inquirers working with big data sets. The authors importantly point to how big data in qualitative research has the potential of allowing for theoretical generalisability in compelling and much needed ways. In this timely book, the authors walk the reader through not just the value and importance of big qualitative data, but also how to access and create such datasets. Perhaps most significantly, this book discusses in great detail the 'breadth-and-depth analysis' approach to working with big qualitative data. This approach offers readers a way to think about how to analyze big qualitative data in meaningful and theoretically grounded ways. Weller et al.'s book is indeed long overdue and will serve as an important guidepost for the qualitative research community as we increasingly encounter and work with big qualitative datasets".

—Jessica Nina Lester, *Professor of Qualitative Methodology,*
Indiana University Bloomington, USA

"*Big Qual: A Guide to Breadth-and-Depth Analysis* is the ultimate guide for learning how to do big qualitative research. The rapidly expanding world of big data calls for new solutions to manage and analyse large volumes of qualitative data now becoming available to social scientists everywhere. Moving well beyond rhetoric, this book takes readers on a journey to discover the newly minted 'breadth-and-depth analysis' method for big qualitative data analysis. It gives easy-to-follow steps, with excursions into data mining tools, and supported by rich, illustrative case studies. *Big Qual* will lead you on a rewarding quest to navigate qualitative data landscapes, from a birds-eye aerial reconnaissance to a detailed charting of the data terrain. Begin to unearth what hides beneath the surface of digital data assemblages, in an iterative journey to uncover and find meaning in a data-saturated social world".

—Kathy A. Mills, *Australian Catholic University, Australia*

Susie Weller • Emma Davidson
Rosalind Edwards • Lynn Jamieson

Big Qual

A Guide to Breadth-and-Depth Analysis

Susie Weller
University of Oxford
Oxford, UK

Rosalind Edwards
SSPC, University of Southampton
Southampton, UK

Emma Davidson
University of Edinburgh
Edinburgh, UK

Lynn Jamieson
Centre for Research on Families and Relationships
University of Edinburgh
Edinburgh, UK

ISBN 978-3-031-36323-8 ISBN 978-3-031-36324-5 (eBook)
https://doi.org/10.1007/978-3-031-36324-5

© The Editor(s) (if applicable) and The Author(s), under exclusive licence to Springer Nature Switzerland AG 2023
This work is subject to copyright. All rights are solely and exclusively licensed by the Publisher, whether the whole or part of the material is concerned, specifically the rights of translation, reprinting, reuse of illustrations, recitation, broadcasting, reproduction on microfilms or in any other physical way, and transmission or information storage and retrieval, electronic adaptation, computer software, or by similar or dissimilar methodology now known or hereafter developed.
The use of general descriptive names, registered names, trademarks, service marks, etc. in this publication does not imply, even in the absence of a specific statement, that such names are exempt from the relevant protective laws and regulations and therefore free for general use.
The publisher, the authors, and the editors are safe to assume that the advice and information in this book are believed to be true and accurate at the date of publication. Neither the publisher nor the authors or the editors give a warranty, expressed or implied, with respect to the material contained herein or for any errors or omissions that may have been made. The publisher remains neutral with regard to jurisdictional claims in published maps and institutional affiliations.

Cover illustration © xia yuan / Getty Images

This Palgrave Macmillan imprint is published by the registered company Springer Nature Switzerland AG.
The registered company address is: Gewerbestrasse 11, 6330 Cham, Switzerland

Paper in this product is recyclable.

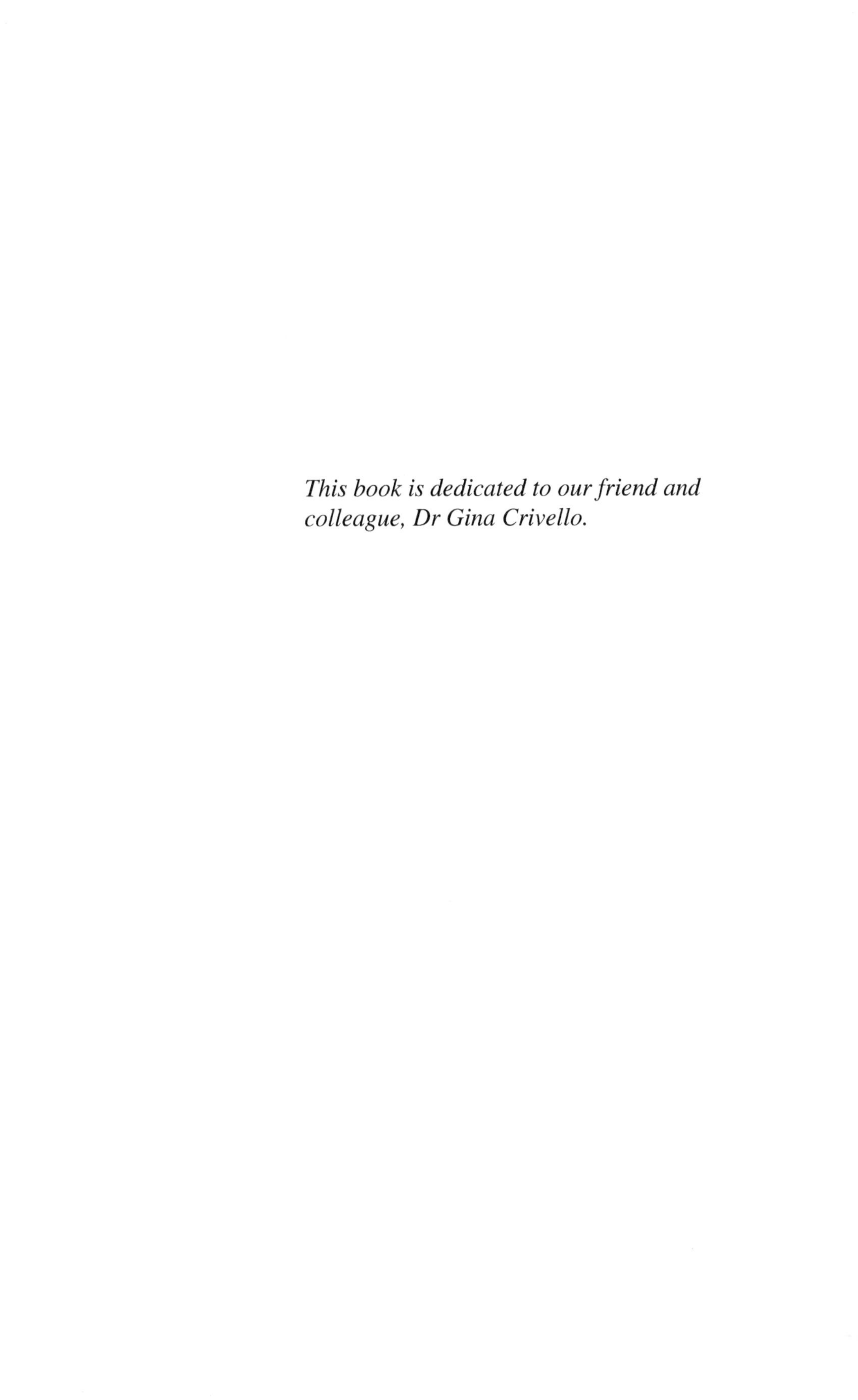

This book is dedicated to our friend and colleague, Dr Gina Crivello.

Foreword

It gives me great pleasure to introduce this landmark publication. In straddling qualitative secondary analysis (QSA) and Big Data analytics, Big Qual represents a quantum leap in the interpretation of qualitative research data.

My interest in QSA was sharpened in 2007, when I began working with a national team of talented researchers on the five-year Economic and Social Research Council (ESRC)-funded Timescapes Initiative. During this time, we completed seven qualitative longitudinal studies that were methodologically and thematically linked. Building on the work of the Qualidata team at the UK Data Archive, we worked with our archivist Libby Bishop and the University of Leeds Data team to draw these wonderfully rich and extensive datasets together to create the Timescapes Data Archive. We were able to explore new ways to digitise and preserve these data and facilitate their discovery through new search and retrieval tools. The re-use of the datasets was showcased in a dedicated QSA project led by Sarah Irwin, working in collaboration with Joanna Bornat. These were exciting times: we were in the vanguard of new developments for scaling up and re-purposing legacy datasets, and creating an infrastructure within which such research could flourish (Neale, 2021).

In the intervening years, QSA has become a popular and well-established approach to social enquiry. The gradual expansion of qualitative data archives and local archival collections, along with a greater acceptance of research data as a public good for sharing and re-use, has created a world of possibilities for reconfiguring and aggregating qualitative datasets across time and space. In 2020, Kahryn Hughes (a Timescapes researcher) and Anna Tarrant (a secondary user of the Timescapes Data Archive) edited a 'state of the art' collection on QSA that reflected these developments and has since inspired further advances (Hughes and Tarrant, 2020).

This brings me to Big Qual. The authors have been working at the forefront of developments in QSA for the past decade and more. Susie, Rosalind and Lynn were core members of the Timescapes team. Building on this foundation, and with the benefit of an 'outsider' perspective from Emma, they have used the rich resources of the Timescapes Data Archive as a test bed for advancing Big Qual, and have taken methods for aggregating, repurposing and analysing qualitative datasets to new and exciting levels of sophistication.

Two distinctive aspects of their approach are worth highlighting here. Firstly, the team has devised ways to merge quantitative (computational) methods of data discovery, searching and sampling with a qualitative framework for interpretation and the production of meaning. They are able to combine an extensive breadth of coverage with an intensive depth of understanding, and, in the process, to blur the distinctions that are commonly drawn between different scales of social enquiry. While their breadth-and-depth method retains a central qualitative focus on narratives rather than numbers, and on meanings rather than measurements, they have also set new agendas for bridging qualitative and quantitative approaches to the secondary analysis of research data.

Secondly, they have helped to establish and advance the fledgling field of Big Qual, an approach that sits firmly within the burgeoning discipline of Big Data analytics. At this point it is worth clarifying what Big Qual actually is (and isn't). This is not a book about Big Data per se (with its connotations of extensive 'thin' data that are only amenable to statistical analysis). Nor is it about ways to work qualitatively with these 'thin' data (although insightful discussions about this approach are included). This book is about how to work with big, thick, qualitative data, that is, very large volumes of narratives, texts, and visual and aural data, which can be drawn together from a variety of institutional and personal archives and other sources to address new research questions. The scale of the enterprise (some studies involve the aggregation of data from over 1000 cases) is such that a solo qualitative researcher or small team would be unable to comprehend, let alone harness, these data using standard manual analytical strategies. Big Qual, then, creates a quantum leap in the scale of QSA.

Working qualitatively at such scale has its own powerful rationale: elaborating and synthesising many pockets of evidence, drawn from a carefully chosen range of contexts and settings, enables the production of qualitative generalisations of a different order of magnitude from standard approaches: it creates a seismic shift in the nature of qualitative evidence and findings. Insights about patterns and processes of change are given added credence through the extent, variety and weight of the evidence base. Moreover, the adoption of abductive (iterative) logic and reasoning, in which the researcher zig zags back and forth between theoretical precepts and empirical evidence, intuition and precision, and depth and breadth, lends further weight to the rigour and integrity of this approach.

The nuts and bolts of the breadth-and-depth method, with its cumulative momentum, is set out here with great clarity. Computational methods are used to map out, search and sample large pockets of qualitative data, while in-depth analytical strategies are employed to identify salient cases, assemble them into new configurations, and draw out their meanings. While computational methods are vital for mapping, sampling and finding ways into these vast stores of data, they have little purchase when it comes to data reconfiguration and analysis, which relies on the analytical and interpretive skills of the researcher.

Detailed discussions of this cumulative and iterative process are interspersed with a range of interesting case studies and exemplars that flesh out the picture and bring the breadth-and-depth method alive. In keeping with the qualitative grounding of Big Qual, the theoretical, epistemological and ethical underpinnings of this approach are also explored. This, then, is far from a 'cookbook' approach to research methods: there is nothing prescriptive about the breadth-and-depth approach, for it is not designed to replace existing methodologies but, rather, to enrich them. The authors do this admirably, creating new and important methodological insights, and a new orientation to QSA and Big Qual that substantially advances both fields.

This brings me to some final reflections. In combining quantitative breadth with qualitative depth in the process of secondary analysis, this book offers a great deal to researchers from both traditions, even though it may take them out of their comfort zones. Qualitative researchers will be reassured to know that using computational methods to work with large volumes of qualitative data doesn't require extensive training; it requires familiarity with text mining computer packages such as Leximancer, which are generally accessible and easy to learn through the training modules provided by this team. More generally, engaging with the scale of the social fabric is vital in a fast moving, increasingly complex and expansive global world. The particularities of qualitative evidence stand on their own merits, but they are enriched when seen as part of this broader picture. At the same time, qualitative researchers have a great deal to offer the burgeoning field of Big Data Analytics: working at scale through collaborative modes of knowledge production can help to ensure that Big Qual finds its place as an integral component of Big Data Analytics.

The same reflections apply to quantitative researchers, whose engagement would take them beyond the solid, comfortable zone of numbers and probabilities, precision and certainty. The value of taking this more open, exploratory approach has been sharpened in recent years. There is a growing awareness of the need to go beyond 'hard data' to develop interpretative modes of enquiry that can better reflect the particularities of lives, and the complexities of causal processes in a social landscape that is endlessly varied, increasingly fluid and inherently unpredictable (Neale, 2021). For both groups of researchers, then, the iteration between breadth and depth becomes vital. In sum, this book provides the perfect starting point for exploring innovative and collaborative modes of enquiry that offer new and exciting ways to know and understand the social world.

University of Leeds, Leeds, UK Bren Neale

About the Contributors

Case Study 3.1: Burundi's Statistical Visa: Coordinating or Controlling Qualitative and Quantitative Data Collection? *Jean-Benoît Falisse, University of Edinburgh, UK; Godefroid Bigirimana, United Nations Development Programme (UNDP), Brussels*

Case Study 3.2: Sharing Qualitative Research Data in Germany with Qualiservice *Jan-Ocko Heuer, Qualiservice, University of Bremen, Germany*

Case Study 3.3: The Timescapes Archive *Kahryn Hughes, Timescapes Archive, University of Leeds, UK; Bren Neale, University of Leeds, UK; Graham Blyth, Data Management team, University of Leeds, UK; Brenda Phillips, Institutional Repository team, University of Leeds, UK; Rachel Proudfoot, Data Management team, University of Leeds, UK*

Case Study 3.4: Working across Data Sets in the Irish Qualitative Data Archive: An Example and Future Possibilities *Jane Gray, Faculty of Social Sciences, Maynooth University, Ireland*

Case Study 3.5: Saving and Reanimating a Classic Data Set: The Women, Risk & AIDS Project *Rachel Thomson, University of Sussex, UK*

Case Study 4.1: Building New Empirical Research using Archived Qualitative Data; 'Men, Poverty and Lifetimes of Care' *Anna Tarrant, University of Lincoln, UK*

Case Study 4.2: How a Team of Researchers Undertook an Initial Exploration of HIV and Biomedicalisation across 12 UK Qualitative Data Sets *Catherine Dodds, School for Policy Studies, University of Bristol, UK.*

Case Study 4.3: Young Lives: Linking Large Volumes of Survey and Qualitative Data across Time *Gina Crivello, University of Oxford, UK*

Case Study 4.4: QSA in the Sexuality and Abortion Study *Carrie Purcell, Open University, UK; Karen Maxwell, Glasgow Caledonian University, UK*

Case Study 5.1: Comparative Keyword Analysis Extract from Seale, C., Ziebland, S. & Charteris-Black, J. (2006) Gender, Cancer Experience and Internet Use: A Comparative Keyword Analysis of Interviews and Online Cancer Support Groups. *Social Science & Medicine,* 62(10), 2577–90 *Clive Seale, Brunel University, UK; Sue Ziebland, University of Oxford, UK; Jonathan Charteris-Black, University of the West of England, UK*

Case Study 5.2: The History of Feminist Ideas to 'Computational Grounded Theory' Summary and extracts from: Nelson, L. (2021). Cycles of Conflict, a Century of Continuity: The Impact of Persistent Place-Based Political Logics on

Social Movement Strategy. *American Journal of Sociology 127*(1), 1-59, 202 and Nelson, L. (2020). Computational Grounded Theory: A Methodological Framework, *Sociological Methods & Research, 49*(1), 3-42 *Laura Nelson, University of British Columbia, Canada*

Case Study 6.1: Analysing Students' Employability Narratives: From Breadth of a Concept Map to Depth of Collective and Individual Narratives. *Elena Zaitseva, Teaching and Learning Academy, Liverpool John Moores University, UK*

Case Study 6.2: Young People and Food Choice in Transition *Mary Barker, MRC Lifecourse Epidemiology Centre, University of Southampton, Southampton General Hospital, UK; School of Health Sciences, Faculty of Environment and Life Sciences, University of Southampton, UK; Polly Hardy-Johnson, Primary Care Population Sciences and Medical Education (PPM), Faculty of Medicine, University of Southampton, UK; MRC Lifecourse Epidemiology Unit, University of Southampton, Southampton General Hospital, UK; Sofia Strömmer, MRC Lifecourse Epidemiology Unit, University of Southampton, Southampton General Hospital, UK; Susie Weller, Clinical Ethics, Law and Society (CELS) and Centre for Personalised Medicine, University of Oxford, UK; National Centre for Research Methods, University of Southampton, UK*

Case Study 6.3: Moving Home across the Life-course *Rosalind Edwards, University of Southampton, UK; Susie Weller, Clinical Ethics, Law and Society (CELS) and Centre for Personalised Medicine, University of Oxford, UK; National Centre for Research Methods, University of Southampton, UK; Lynn Jamieson, University of Edinburgh, UK; Emma Davidson, University of Edinburgh, UK*

Case Study 7.1: The Possibilities of Paradata: An Historically Situated Exploration of 'BigQual' Marginalia *Ann Phoenix, Institute of Education, University College London, UK; Heather Elliott, University of East London, UK; Janet Boddy, University of Sussex, UK; Rosalind Edwards, University of Southampton, UK*

Case Study 7.2: Growing Up on the Streets *Lorraine van Blerk, University of Dundee, UK; Janine Hunter, University of Dundee, UK; Wayne Shand, Consultant at EDP Associates Ltd., UK*

Case Study 7.3: Abductive Coding: The QUALIDEM Project *Claire Dupuy, University of Louvain, Belgium; Virginie Van Ingelgom, University of Louvain, Belgium; Luis Vila-Henninger, Aarhus University, Denmark*

Case Study 8.1: Collaborating with the Original Research Teams *Susie Weller, Clinical Ethics, Law and Society (CELS) and Centre for Personalised Medicine, University of Oxford, UK; National Centre for Research Methods, University of Southampton, UK*

Case Study 8.2: Everyday Childhoods *Liam Berriman, University of Sussex, UK*

Case Study 8.3: Linked Lives and Linked Data: A QLR Study of Journeys through Genomic Medicine *Susie Weller, Clinical Ethics, Law and Society (CELS) and the Centre for Personalised Medicine, University of Oxford, UK; National Centre for Research Methods, University of Southampton, UK; Kate Lyle, Clinical Ethics, Law and Society (CELS), University of Oxford, UK; Anneke Lucassen, Clinical Ethics, Law and Society (CELS) and the Centre for Personalised Medicine, University of Oxford, UK*

Case Study 8.4: Shifting Connections to Data in the Timescapes Qualitative Longitudinal Archive *Susie Weller, Clinical Ethics, Law and Society (CELS) and the Centre for Personalised Medicine, University of Oxford, UK; National Centre for Research Methods, University of Southampton, UK; Rosalind Edwards, University of Southampton, UK*

Preface

Internationally, the number and scale of qualitative data sets available for reuse has grown, a development supported by archiving, data digitisation, and other forms of data storing and sharing. This presents new and exciting opportunities for qualitative research, as demonstrated by the burgeoning field of qualitative secondary data analysis (Hughes and Tarrant, 2020). While qualitative forms of enquiry are expanding, there is yet to be fulsome engagement in the opportunities that can be gained from analysing very large volumes of qualitative data, otherwise known as 'big qual'. Here analytical techniques remain in their infancy, often relying on a descriptive content analysis of words, themes, and concepts, or on efforts to quantify qualitative data. This can result in the loss of rich, in-depth understandings central to qualitative forms of enquiry.

In this book, we present the breadth-and-depth method, a new iterative approach to managing and analysing large volumes of qualitative data. By combining advances in computational text analysis (the breadth) and qualitative data analysis (the depth), we present an approach for 'pooling' multiple qualitative data into a new data assemblage. Bringing data sets together in this way increases the diversity of sample populations and contexts, enhances the possibility of theoretical generalisability, and, in turn, strengthens claims from qualitative research about understanding how social processes work.

The approach resulted from an ESRC National Centre for Research Methods methodological project, 'Working across qualitative longitudinal studies: a feasibility study looking at care and intimacy' (http://bigqlr.ncrm.ac.uk/). That project ran for four years (2015–2019) and examined the possibilities of developing new procedures for working across multiple sets of archived qualitative data.

The project and this book are a response to two interconnected methodological developments. The first is the rapidly expanding open science movement and associated drive to promote data sharing across the humanities and social sciences. Publicly funded agencies are increasingly placing requirements on researchers to make their original data accessible to the wider research community. While such requirements are now fairly commonplace for quantitative researchers, open access continues to pose epistemological and methodological questions for qualitative researchers. In this context, archives, supported by new information systems and digital technologies, have emerged both as a critical infrastructure for storing qualitative research data and a source of good practice for qualitative researchers wishing to engage in sharing and reuse. This movement continues to expand internationally, reflected in the growing numbers of initiatives and projects re-using qualitative data.

The second associated development is the expansion of computational text analysis, which has provided researchers with a range of sophisticated techniques for analysing enormous amounts of data speedily. Such tools and techniques have advanced as part of the 'big data movement', with the central aim of exploiting the massive amounts of data available from various sources and turning it into meaningful data. However, erroneously, computational text analysis and the associated tools can prioritise quantitative knowledge, and lack the necessary theoretical framing and rigour in interpretation for dealing with qualitative data sources. It is only in recent years that scholars have begun to explore in detail the potential of big data and big data analytics in unpacking the qualitative dimensions of digitally born data.

While big qual analysis has begun to develop and expand, these developments raise the prospect and theoretical importance of a new form of big qualitative analysis, one that involves merging qualitative data from several discrete studies. Our methodological project, upon which this book is based, was an attempt to create an approach for conducting big qual analysis in a way that retained the distinctive characteristics of rigorous qualitative research. We did not wish to use computational techniques uncritically at the expense of the central tenets of qualitative work. Concurrently, we wanted to bring big qual into wider debates by developing a form of analysis where computational processing of large volumes of qualitative data is given equal prominence to 'deep data' research approaches. The desire was to draw on the new techniques afforded by big data analytics to expand qualitative understandings.

When we began our project, we became aware of the limited tools and guidance available to support researchers—across the qualitative–quantitative spectrum—in navigating projects involving large volumes of qualitative data. There were few examples of projects systematically merging multiple qualitative data sets. Our aim, therefore, was to create a step-by-step process that researchers could apply flexibly whatever the theoretical logic, substantive topic, and nature of the qualitative data. Our strategy revolved around addressing three key questions:

- How feasible is it to conduct secondary data analysis across existing data from multiple qualitative studies?
- What is the relationship between breadth of analysis and depth of analysis in working with large and/or across multiple data sets?
- Can we take a 'test pit' approach to trying out different complementary forms of qualitative data analysis as a means of selecting cases for full analysis?

We developed the method using the first major qualitative longitudinal study to be funded in the UK, the Timescapes initiative (https://timescapes-archive.leeds.ac.uk/). Across several separate but connected research projects, Timescapes explored how personal and family relationships develop and change over time. It ran for five years from February 2007 and was funded by the Economic and Social Research Council (ESRC) under their Changing Lives and Times initiative. The broad aim was to scale up and promote Qualitative Longitudinal Research (QLR), create an archive of data for preservation and sharing, and demonstrate and encourage reuse of the resource through a range of training and capacity building activities (Holland and Edwards, 2014, Neale and Bishop, 2012).

Our interest in extending secondary analytic practice originated from our previous methodological qualitative and QLR work and our substantive interest in the life course, care, and intimacy (see, e.g., Edwards et al. 2021a). As a team we were involved in different capacities with the Timescapes initiative, and collectively we were aware of the wealth of information amassed and archived. Despite some variation in substantive focus and method, the projects all traced personal and family relationships over time, each emphasising a different element of the life course. Not only did Timescapes provide exemplars of coherently archived data sets, but it was also suitable for our project's overall focus on care and intimacy practices. Six empirical projects, archived and made available digitally for reuse through Timescapes, were selected as sources of data for our methodological project (see Table 1).

Table 1 Archived data sets used to develop the breadth-and-depth method

Project	Dataset citation
Siblings and Friends	Edwards, R. and Weller, S. (2011) Siblings and Friends: The Changing Nature of Children's Lateral Relationships Dataset. University of Leeds, UK: Timescapes Data Archive. DOI: 10.23635/07
Dynamics of Motherhood	Hadfield, L., Kehily, J., Thomson, R. and Sharpe, S. (2011) Dynamics of Motherhood Dataset. University of Leeds, UK: Timescapes Data Archive. DOI: 10.23635/01
Men as Fathers	Henwood, K., Shirani, F. and Coltart, C. (2011) Masculinities, Identities and Risk: Transition in the Lives of Men as Fathers Dataset. University of Leeds, UK: Timescapes Data Archive. DOI: 10.23635/06
Work and Family Lives	Cunningham-Burley, S., Jamieson, L. and Harden, J. (2011) Work and Family Lives Dataset. University of Leeds, UK: Timescapes Data Archive. DOI: 10.23635/03
Intergenerational Exchange: Grandparents, Social Exclusion and Health	Hughes, K. and Emmel, N.D Access to the Intergenerational Exchange dataset is by special permission only, because it has not been anonymised; it focuses on experiences of place, making the participants highly identifiable, and contains sensitive data.
The Oldest Generation	Bornat, J. and Bytheway, B (2011) The Oldest Generation: Events, Relationships and Identities in Later Life Dataset. University of Leeds, UK: Timescapes Data Archive. DOI: 10.23635/02

How to Use This Book

The chapters of this book are organised around the broad stages of the breadth-and-depth method and explain how the next generation of social scientists can embrace big qual through combining computational text analysis with rigorous in-depth qualitative methods. Each chapter can be followed as an independent module, with diverse case studies demonstrating the rich canon of big qual research taking place internationally. We do, however, wish to emphasise the iterative nature of the breadth-and-depth method. Such analytic work relies not on sequential movement, but rather a back and forth, cyclical movement between the stages. We suggest you bear this in mind as you navigate the pages of the book. Overall, we aim to provide a guide to thinking through the possibilities of big qual and its relationship to qualitative research, as well as affordances for quantitative researchers. In so doing, we prompt readers to situate big qual in a wider context and encourage engagement in the relationship between theory and evidence, research questions and data.

The book is divided into four parts. Part I is concerned with setting out the context of big qual data analysis and its relationship to the wider big data movement. In Chap. 1 we set the scene by discussing the turn to big data and reviewing the emergent work that critically examines the epistemological challenges it brings to qualitative research. Crucially, we set out the new big qual field, and the diverse range of research taking place. Chapter 2 then introduces readers to the breadth-and-depth method and the unique contribution it makes to these wider methodological developments.

Part II moves on to the first stages of the breadth-and-depth method, with a focus on searching, sourcing, and surveying your selected data. Chapter 3 begins by introducing readers to the expanse of possible data sources appropriate for your study, while Chap. 4 deals with techniques and approaches that can be used to gain an overview of your selected data sets, as well as managing the challenges of constructing a new corpus.

In Part III we begin to demonstrate how, through an approach of qualitative layering, the method moves iteratively between breadth and depth. Chapter 5 provides a comprehensive overview of data mining tools and their role in thematically mapping your data to identify areas suitable for further exploration. Chapter 6 moves on to demonstrate how these points of interest can be subjected to preliminary analysis, and then in Chap. 7 we detail how your final selection of data cases can be analysed using a range of qualitative data analysis techniques.

Part IV provides a final section in which we explore the wealth of possibilities and opportunities that can stem from the breadth-and-depth method, as well as the challenges. While ethics is a continuous theme throughout the book, in Chap. 8 we provide rich reflection on the ethical implications of big qual analysis. This covers issues such as the ethical practices associated with constructing a corpus, how to work with qualitative integrity at scale, and the issues associated with working qualitatively with data analytical tools. If you have chosen to read only specific chapters of the book, we encourage you to include this chapter in your reading also. Our final chapter zooms back out, taking a final reflective look at big qual and the future of qualitative analysis. The focus here, and throughout the book, is on supporting researchers to use the breadth-and-depth method and contribute to expanding the qualitative endeavour in new and exciting directions.

Resources

The Timescapes Archive is a specialist resource of Qualitative Longitudinal Research [QLR] data: https://timescapes-archive.leeds.ac.uk/

This podcast provides a good overview of the Timescapes Archive: https://www.ncrm.ac.uk/resources/podcasts/mp3/NCRM_podcast_timescapes.mp3

This podcast provides a useful introduction to the breadth-and-depth method:

Weller, S. (2017). *Digging deep! The archaeological metaphor helping researchers get into big qual*. https://www.ncrm.ac.uk/resources/podcasts/mp3/NCRM_podcast_Weller2.mp3

Key readings include:

Davidson, E., Edwards, R. Jamieson, L., & Weller, S. (2019). Big data, qualitative style: A breadth-and-depth method for working with large amounts of secondary qualitative data. *Quality & Quantity, 53*(1), 363–376.

Edwards, R., Davidson, E., Jamieson, L., & Weller, S. (2021). Theory and the breadth-and-depth method of analysing large amounts of qualitative data: A research note. *Quality & Quantity, 55*, 1275–1280.

Weller, S., Davidson, E., Edwards, R., & Jamieson, L. (2019). *Analysing large volumes of complex qualitative data: Reflections from international experts*. NCRM Working Paper. https://eprints.ncrm.ac.uk/id/eprint/4266/

Lewthwaite, S., Weller, S., Jamieson, L., Edwards, R., & Nind, M. (2019). *Developing pedagogy for Big Qual Methods: Teaching how to analyse large volumes of secondary qualitative data*. NCRM Working Paper. https://eprints.ncrm.ac.uk/id/eprint/4247/

References

Holland, J., & Edwards, R. (2014). Introduction to Timescapes: Changing relationships and identities over the life course. In J. Holland, & R. Edwards (Eds.), *Understanding families over time: Research and Policy* (pp. 1–28). Palgrave Macmillan UK.

Hughes, K., & Tarrant, A. (2020). *Qualitative secondary analysis*. Sage.

Neale, B., & Bishop, L. (2012). The timescapes archive: A stakeholder approach to archiving qualitative longitudinal data'. *Qualitative Research, 12*, 53–65.

Acknowledgements

This book would not have happened were it not for the ESRC National Centre for Research Methods (NCRM) who in 2015 funded what was to be a four-year feasibility study on analysing large volumes of qualitative data. With their support and encouragement, the project became much more than initially planned, and over the years has generated extensive outputs. This book is the culmination of this work.

We, of course, must also extend our thanks to those involved in the Timescapes Initiative past and present. Without being familiar with the research data from this wonderful collection of qualitative longitudinal studies, our project would never have started. Throughout the study, we consulted with researchers involved in studies across all stages of the initiative. All were generous with their time, willingly providing advice, support, and guidance. This support helped us to contextualise the data we were working with, but also fill gaps in our theoretical thinking and navigate ethical conundrums. This project also led to additional funding which explored teaching approaches to big qual with Melanie Nind and Sarah Lewthwaite. Special thanks to all those involved, especially Joanna Bornat, Nick Emmel, Kahryn Hughes, Bren Neale, Brenda Phillips, Anna Tarrant, Rachel Thomson, and the Timescapes research participants. And of course, to Justin Chun-ting Ho, who supported the team in our learning about computational methods.

We would like to thank all those who supported our endeavour by contributing rich and insightful case studies: Mary Barker, Liam Berriman, Godefroid Bigirimana, Graham Blythe, Janet Boddy, Gina Crivello, Catherine Dodds, Claire Dupuy, Heather Elliott, Jean-Benoît Falisse, Jane Gray, Polly Hardy-Johnson, Jan-Ocko Heuer, Kahryn Hughes, Janine Hunter, Anneke Lucassen, Kate Lyle, Bren Neale, Brenda Phillips, Ann Phoenix, Rachel Proudfoot, Wayne Shand, Sofia Strömmer, Anna Tarrant, Rachel Thomson, Lorraine van Blerk, Virginie Van Ingelgom, Luis Vila-Henninger, and Elena Zaitseva.

We would especially like to thank Alan Marshall and Michael Rosie, who gave their time and expertise so generously in reviewing the manuscript.

We would also like to thank the team at Palgrave Macmillan for supporting us in the development of the manuscript, especially through the pandemic when our caring and other responsibilities impacted on our lives in such unexpected ways.

Funding

This work was supported under the five-year umbrella grant for the Economic and Social Research Council National Centre for Research Methods (Grant ID: ES/L008351/1).

About the Book

When social scientists think about big data, they often think in terms of quantitative data sets that can be analysed computationally to reveal patterns, trends, and associations, especially relating to human behaviour and interactions. However, there exists a growing wealth of qualitative data sets, available through archives and other forms of data sharing. These data present new and exciting opportunities internationally for qualitative research. Pooling multiple qualitative data sets enhances the possibility of theoretical generalisability by increasing the diversity of sample populations and contexts. This can, in turn, strengthen claims from qualitative research about how social processes work.

In *Big Qual: A Guide to Breadth-and-Depth Analysis* we present a new approach for working with large volumes of qualitative data. As the name suggests, the method brings together breadth and depth through an integration of computational and qualitative data analysis. This book covers everything researchers—both qualitative and quantitative— need to know about analysing large volumes of qualitative data, from sourcing and creating your data set, to applying the method, to understanding the ethical and epistemological challenges.

After presenting a comprehensive overview of the rationale for, and value of, large-scale qualitative data analysis, the book takes readers through four stages of dealing with big qual that are akin to the process of discovery in archaeology:

- An enquiry-led overview of archived qualitative research: using metadata like an archaeologist uses photographs in an aerial survey
- Computer-aided scrutiny across the breadth of selected data collections: to assess what merits closer investigation, like an archaeologist's ground-based geophysical survey of an area
- Analysis of multiple small samples of likely data: equivalent to digging shallow 'test pits' to find an area meriting deeper excavation.
- In-depth analysis: of the type familiar to qualitative researchers, like archaeological deep excavation

The aim is not to create a prescriptive method or methodology for students, researchers, and teachers of research methods to deploy. Rather, the book provides a guide to thinking through the possibilities of big qual and its relationship to qualitative research more generally. The breadth-and-depth approach can also function in a modular fashion, with each chapter of the book prompting readers to think differently about the relationship between theory and evidence, research questions and data. Combined with tried and tested linked multimedia resources, expert international case studies and a bespoke teaching data set, the book is a unique addition to textbooks seeking to analyse qualitative data using computational methods.

This book is intended for international audiences with a stake or interest in the use of qualitative data, regardless of discipline and methodological background. Since our approach meshes quantitative and qualitative approaches, we hope that it can be used to encourage collaborative and cross-disciplinary thinking. Given the continued rise of big data and the evolving possibilities it offers the humanities and social sciences, we hope that it will build capacity in the UK and internationally in 'big qual' research, and provoke new and exciting ways of thinking about, and working with, large volumes of qualitative data and its analysis.

Contents

About the Authors

Susie Weller (she/her) has 20 years' experience of conducting research with children, youth, and families. She has been principal investigator, co-investigator, or senior researcher on a range of qualitative longitudinal studies examining connections between individual experiences and wider social processes. Throughout her career, she has worked at the interface between theoretical advances in youth and family research and applied research that has policy and practice relevance. Much of this work has been in interdisciplinary teams straddling the social and biomedical sciences. Susie has (co)authored over 75 papers published in peer-reviewed journals, books, and working papers including in *The Lancet, Social Science and Medicine, the Journal of Medical Ethics, BMC Public Health, Qualitative Research, Sociological Research Online,* and the *International Journal of Social Research Methodology.* She is a keen methodologist with expertise in qualitative longitudinal research, creative and participatory approaches, archiving, and the ethical reuse of data. With colleagues, she pioneered the breadth-and-depth method of analysis and co-founded the Big Qual Analysis Resource Hub (http://bigqlr.ncrm.ac.uk/). She has collaborated with cultural industries, policy makers, practitioners, and charities.

Emma Davidson (she/her) is a sociologist with a background in qualitative policy analysis, research, and teaching. A director of the Centre for Research on Families and Relationships, she is passionate about sharing expertise in qualitative research methods through teaching and knowledge exchange. Emma's substantive research is concerned with the relationship between macro socio-economic structures and the subjective, micro-personal. She draws on qualitative methods, typically ethnographic approaches, that help understand people and their lives from their own vantage points. Her areas of research include evaluating the impact of book-sharing; analysing the impact of interventions to address multiple needs homelessness; young people's experiences of antisocial behaviour policy; the role of informal support in the care system; and interrogating evidence on the benefits of Adverse Childhood Experiences-informed policy. Most recently, Emma has been leading research funded by the Leverhulme Trust on the everyday social world of the public library and the challenges faced under austerity. Emma's work has been published in *Social Policy & Society, Social Policy Review, Scottish Affairs, Social Inclusion, Sociological Research Online* and *Quality and Quantity.*

Rosalind Edwards (she/her) has researched and published widely on qualitative longitudinal research methods, and is a founding editor of the *International Journal of Social Research Methodology*. Her relevant publications include *Understanding Families Over Time: Research and Policy* (Palgrave Macmillan, 2014) and *Researching Families and Communities: Social and Generational Change* (2008). Rosalind's research interests are centred on how to think about and conceptualise family lives; how best to conduct studies into family lives; and what goes on in people's family lives and the ideas and assumptions shaping family policies. She takes a critical sociological approach coupled with feminist relational perspectives to address and understand theory, methods, and substance, and is especially interested in how these are shaped by gender, social class, race/ethnicity, and generation, within geographical, political, and historical contexts.

Lynn Jamieson (she/her) is a professor of the Sociology of Families and Relationships at the University of Edinburgh. She is known for her research on intimacy, identity, and social change and has researched a wide range of topics within the field of families and relationships. Her current research interests include the theoretical significance of people living alone and the power of families and relationships to influence responses to climate change and issues of sustainability. Her portfolio of research also includes work on European identity and gender violence. Lynn was President of the British Sociological Association and is co-editor of the Palgrave Macmillan series Families and Intimate Life and Associate Editor of the Policy Press journal *Families Relationships and Societies*. She is a founding co-director of the Centre for Research on Families and Relationships, a consortium centre established in 2001 to grow capacity for research on families and relationships in Scotland, acting as a hub of a network of knowledge exchange and international collaborations. Her publications include *Intimacy: Personal Relationships in Modern Societies* (1997) and *Researching Families and Relationships: Reflections on Process* (co-edited with Caroline King, Zhong Eric Chen, and Dr Roona Simpson; Palgrave Macmillan 2011).

Abbreviations

ADA	Australian Data Archive
ALLF	Archive for Life Course Research, Germany
CESSDA	Consortium of European Social Sciences Data Archives
CRC	Collaborative Research Center, Germany
CSA	Centre for Scientific Archives
CWLU	Chicago Women's Liberation Union Centre for Scientific Archives
DDA	Danish Data Archive
DRI	Digital Repository of Ireland
ESRC	Economic and Social Research Council, UK
FAIR	Findable, Accessible, Interoperable, and Reusable
FID SKA	Scientific Information Service Social and Cultural Anthropology, Germany
FSD	Finnish Social Science Data Archive
GALA	Gay and Lesbian Memory in Action
GUI	Growing Up in Ireland
HASSET	The Humanities and Social Science Electronic Thesaurus
HE	higher education
ICPSR	Inter-university Consortium for Political and Social Research, USA
IQDA	Irish Qualitative Data Archive
KWIC	keyword in context
LDA	Latent Dirichlet Allocation
LGBTIQ	Lesbian, gay, bisexual, transgender, and queer or questioning
LHSC	Life Histories and Social Change, Ireland
NCRM	National Centre for Research Methods, UK
NCUACS	British National Cataloguing Unit for the Archives of Contemporary Scientists
NFDI	National Research Data Infrastructure, Germany
NIH	National Institutes of Health
NIQA	The Northern Ireland Qualitative Archive
NSS	National Student Survey, UK
PinUK	Poverty in the United Kingdom 1967/68
QLR	Qualitative longitudinal research
QSA	Qualitative secondary analysis
RAKE	Rapid Automatic Keyword Extraction

RatSWD	German Data Forum, Germany
RDC	Research Data Center, Germany
SPSS	Statistical Package for the Social Sciences
SuUB	State and University Library Bremen, Germany
UKDA	UK Data Archive
UKDS	UK Data Service
WISDOM	Wiener Institute for Social Science Data Documentation and Methods

List of Figures

List of Tables

Part I

From Big Qual to the Breadth-and-Depth Method

The Place and Value of Large-Scale Qualitative Data Analysis

1

1.1 Introduction

Big data is not just about numbers. The digitised stories and images that are now routinely produced when lives are lived online, alongside a myriad of more formal digitised documents of life, are generating large and increasingly complex volumes of textual, oral, and visual material. In addition, there is an accumulation of digitised qualitative data that has been assembled by trained academics in the course of their social research. These data are more easily navigable, particularly when they have been systematically curated in archives that are purposefully constructed to enable future reuse. The wealth of digitised qualitative data, but archived qualitative research data in particular, is giving social researchers the opportunity to access new sources, methodological approaches, and forms of knowledge. Big data analytics has typically relied on algorithmic or computational procedures, and despite the accumulation of digital qualitative material, is often associated with the quantification of data. The distinctive skills of the qualitative researcher, which focus on context, subjectivities, and thick understanding, are overlooked in popular understandings of 'big data'.

Bringing qualitative research skills into the burgeoning use of big data in research has the potential to make a transformative impact. Actively integrating qualitative philosophies and analytical techniques can provide balance to more reductive understandings of big data. It can also productively extend and expand qualitative research methods. This is being demonstrated through 'big qual', a field which has developed to include a diverse range of combinations crossing between working qualitatively and quantitatively—quali-quan methods, as well as innovative methods for collecting and analysing large volumes of qualitative data. These approaches are generating integrative spaces, where methodological bridges can be built, not re-ascribed (boyd & Crawford, 2011, 2012).

In this chapter, we review the place and value of these developments within the growing use of big data. We do this so as to situate the breadth-and-depth method and highlight its own unique contribution. We will:

© The Author(s), under exclusive license to Springer Nature Switzerland AG 2023

S. Weller et al., *Big Qual*, https://doi.org/10.1007/978-3-031-36324-5_1

- Review the big data movement and its relevance to qualitative forms of enquiry
- Map the methods and approaches seeking to integrate big data and qualitative forms of enquiry
- Consider future challenges for the big qual field and the contribution of the breadth-and-depth method

1.2 What Is Big Data?

We begin by discussing the meaning of big data and the orientation to its analysis. Taking this as our starting point is important because it is from there that big qual is developing as a field in its own right, both theoretically and in practice. At a simple level, big data is "rapidly generated, digitally encoded information of significant volume, velocity, variety, value and veracity" (Mills, 2019, p. 10). This is not simply large volumes of complex, quantitative data, but rather it is digital data created at pace, over time, and not previously available to social scientists. Playford et al. (2016) note the complexity of defining big data and its frequent and erroneous association to size or to one particular form of data. Big data sources are, in fact, multifarious, being both structured and unstructured, and they can comprise numerical data, as well as text, visuals, and audio. They can also span different analytical scales, from information created through mobile and internet-enabled devices or wearable technology, through to transactional and business information, communications, administrative data, as well as CCTV and satellite imaging.

Big data is also tied to data analytics and the "capacity to search, aggregate, and cross-reference large data sets" (boyd & Crawford, 2012, p. 662). Big data, in other words, is intimately related to processes and method. The volume being considered here is so significant that ordinary computing devices are unable to process the complexity of the data sets involved. Volume, therefore, comes hand in hand with the advanced computational techniques necessary to cleanse data, extract it, and make it meaningful. With computational scientists at the helm, the 'big data revolution' has been responsible for driving forward new data architecture, storage, and analytic techniques. What is notable is that a significant amount of big data is, in fact, not quantitative, but textual. Researchers have responded through greater investment in computational text analyses techniques that seek to interrogate the content, patterns, and themes of text with ever greater speed and accuracy.

The attention given to new analytic technologies and techniques reflects the social, economic, and cultural value accorded to big data. As Mills (2019, p. 10) notes, this is data "used as valued evidence of a phenomena". Critically for business, industry, and state agencies, it has brought the tantalising possibility of data-driven insight into the thoughts and behaviours of the general public(s) (Kitchin & Mcardle, 2016). Yet the well-trodden, popular belief that 'big' provides 'better' or more valuable data (Mayer-Schönberger, 2013) overlooks more fundamental consequences. Big data's proliferation should not only be of concern to those in business or marketing. It has altered and continues to alter how we work as social researchers. It has brought:

> ... a profound change [to research] at the levels of epistemology and ethics. Big data reframes key questions about the constitution of knowledge, the processes of research, how we should engage with information, and the nature and the categorization of reality ... big data stakes out new terrains of objects, methods of knowing, and definitions of social life. (boyd & Crawford, 2012, p. 665)

Big data has undoubtedly played a part in shaping ways of thinking and knowing about social relations, networks, and interactions in a complex, globalised world. New technologies, that multitude of ever-developing processing tools and procedures, are in turn contributing to changes in methodological regimes and ways of producing social knowledge. The "computational turn" (Berry, 2011, p. 1465) is working generatively, with the ever-growing presence of data serving to extend the techniques and theories we can use, and to shape the research questions we want and are able to ask.

1.3 Big Data and Qualitative Research

Let us turn now to big data and qualitative research. While big data has a number of characteristics (volume, velocity, variety, and veracity), it is its size and concern with data quantification that has placed it in opposition to the ontological and epistemological foundations of qualitative research. Qualitative research has a variety of histories (see Edwards & Holland, 2013), although it has typically encompassed heterogeneous forms of data, variously text, images, and/or oral recordings, and epistemological positions that require attention to the social context in which data is produced and analysed. Most qualitative research endeavours involve generating new data, although researchers are increasingly drawing on data sets created by previous research projects, as well as structured and unstructured data from other sources, such as archives and online materials.

Through multiple systems of enquiry (discussed further in Chap. 7), all forms of qualitative research seek in-depth understandings of how an individual subjectively perceives and gives meaning to their social reality. The focus, therefore, is on interrogating and exploring the 'what', 'how', and 'why' of social phenomena or behaviour. The analytical work of a qualitative researcher involves familiarisation and immersion variously gained through multiple readings and listenings; precis and memoing; and identifying narratives, themes, meanings, and processes in the data. And because of the time required for rich data collection, deep readings, and nuanced, contextualised analysis, as well as a process-oriented understanding of generalisability, qualitative sample sizes tend to be small.

While what constitutes 'small' is contextual and can vary from project to project, there is an intuitive understanding that qualitative analytical processes limit sample sizes so they are well below 100 (often smaller than ten). For Mason (2012, p. 29), the answer to how big (or small) a qualitative sample should be is 'it depends', and she calls instead for an interpretative and investigative logic which builds a convincing analytical narrative based on "richness, complexity and detail". Theoretical wrangling over qualitative sample size sufficiency is nevertheless commonplace (Braun & Clarke, 2016), and it is not uncommon for qualitative researchers to still find themselves justifying their sample size or troubling over the question 'How many interviews is enough?' (Baker & Edwards, 2012). Bisel et al. (2014, p. 628) note that there is still a tendency to use numerical and quantitative measures to determine 'bigness' in qualitative research. Typically, the question being asked is 'How big was your sample?' For Bisel et al. (2014), more recognition should be given to alternative understandings of small and big. For example, an oral history project might comprise three interviews over multiple conversations. Such a project may only include a few individuals but involve significant emotional and temporal investment in terms of the time spent with participants, the number of words spoken and transcribed, and the complexity of the data. What constitutes big is similarly unfixed. Quantitative researchers, for example, are not only concerned with sample size, but rather give more focus to the size of the sampling error around estimates. Similarly, while more complex modelling techniques rely on large volumes of data, big is not necessarily considered 'better', and disagreement remains over whether more complex quantitative models have greater utility.

This being said, the data sets involved in qualitative studies are certainly nowhere near the terabytes of data involved in much big data analytics. Those working within the emergent field of big qual are still handling, relatively speaking, small data sets in terms of the number of interviews. Brower et al. (2019) have suggested that for a data set to be considered big qual, it should contain either secondary qualitative or primary data with at least 100 participants (Brower et al., 2019, p. 2). Yet computational scientists working routinely with many terabytes may consider such a data set small in comparison. Large qualitative projects working with structured data across several sites may also view 100 interviews as small, relatively speaking. Our own working definition of big qual data analysis does not give a set or rigid numerical value to big data. Rather, we see it as being the analysis of volumes of qualitative data that are much larger than the quantity that would be feasible for a solo researcher or small team to collect and analyse themselves. The forms of data that fall into this definition can then be subjectively determined by researchers themselves, with variations depending on the type and form of investigation.

We will return to discussing big qual research later in the chapter, but for now it is worth reflecting on what the advent of big data and associated tools and techniques means for qualitative research, and whether there is potential for it to be marginalised on the adage that 'bigger' will offer 'better' forms of knowledge or understandings (boyd & Crawford, 2011; Smith, 2014). Kitchin (2014, p. 3) notes that proponents of big data see it ushering in a "new age of empiricism", where the sheer size of the information being generated, coupled with new analytical techniques, can "enable data to speak for themselves free of theory". Big data, in these terms, becomes an unmediated form of knowing and renders the "scientific method obsolete" (Anderson, 2008: np). Kitchin (2014) has problematised this powerful framing of computational research as neutral, free of human bias, and therefore 'better' able to come to a 'true' position about an identified pattern of behaviour. He cautions against such a reductive stance, noting that an individual's behaviour is complex, and cannot be reduced to a computational model.

A related concern is that the volume and procedures associated with big data mean that it operates at "analytical distance" from the data (Harford, 2014, p. 187). This can result in socio-cultural context, meaning, and qualities within being lost, misunderstood, or left unexplained (boyd & Crawford, 2012; Lupton, 2015; Andrejevic, 2014; Marche, 2012). Computer programs, machine learning algorithms, and the insights derived from them are themselves shaped by existing knowledge and the individual subjectivities of the person designing the programs or doing the analysis (Kitchin, 2014). Furthermore, despite advances, computers cannot 'make sense' of data in a way that accounts for subtleties in social context and human behaviour (Silver, 2012). As several commentators in the field note, research which "relies heavily on computational tools for the calculation of large amounts of data and the visualization of patterns still requires the researcher to interpret these patterns" (Van Es et al., 2017, p. 173).

1.4 Breaking Methodological Borderlines

We have spent some time outlining the differences between big data and qualitative research. Pinpointing these differences is important, but we are aware that they might also contribute to boyd and Crawford's (2011, p. 5) suggestion that big data might widen the methodological divide between qualitatively and quantitatively oriented social scientists, or that it might simply be fundamentally resistant to qualitative approaches (Drucker, 2012, p. 86). Interestingly, a further divide might also be open between computational social science and social statisticians who value theoretically informed data collection, such as surveys and censuses, involving probability sampling and theory-led analysis.

It might also support deeper critiques of big data suggesting that it can drive a "mechanistic, atomizing, parochial vision of the world that advances particular kinds of knowledge through elitist, privileged constructions of scientific method" (Wyly, 2014, p. 27). Leurs and Shepherd (2017) have similarly examined discrimination within processes of datafication, outlining concerns that computational practices, rather than being unbiased, can shape and reproduce structural inequalities (see also, e.g., Benjamin, 2019).

Qualitative researchers, in the face of such a powerful framing, might well be forgiven for returning to 'conventional' forms of qualitative data collection and leaving big data well alone. Yet our position is that oppositions and methodological dualisms are not only unhelpful (Rieder & Röhle, 2017), but they fail to recognise longstanding alternative positions. There is—without question—as much 'craft' associated with quantitative work as there is qualitative. Similarly, qualitative research is equally able to come from a positivist tradition as quantitative research. It is also the case that thinking about and attempting to break methodological borderlines is not new. As far back as the 1950s, Kracauer's (1952) work on content analysis argued against the reliance on quantitative analysis, which he asserted not only led to the neglect of qualitative exploration, but overemphasised (incorrectly) the accuracy of quantitative analysis. He concluded that despite assumptions otherwise, the approaches were in fact overlapping and interconnected. More recently, approaches to quali-quantitative research have emerged which seek to meaningfully integrate methods. Elliot, for example, has developed approaches to analysing quantitative data narratively (Elliot, 2005). Rieder and Röhle's (2017) work on the need for 'bildung' is relevant here, where priority is not given to those with computational skills, but rather the ability and willingness to engage in reflexive and collaborative thinking. Garcia and Ramirez (2021, p. 240) use the metaphor of 'a methodological borderland' that embraces the "messiness of theoretical, epistemological, and methodological incompatibility".

Big data raises longstanding epistemological questions, and these require serious, critical attention. The challenge is in finding ways to engage meaningfully with big data and broaden the possibilities it might offer. Such broadening should, we argue, include the integration of big data with qualitative philosophies and approaches. Kitchin's (2014) work is helpful here as it sets out the fallacies of empiricism, drawing the conclusion that no data can speak for itself. Whether quantification or qualitative (or something else entirely), creation will always have context, just as its analysis will always be part of a process of subjective meaning making and interpretation. Suggesting big data is about data points and the quantification of observations and qualitative data is about rich, storied accounts makes methodological assumptions (Wang, 2016). In our own search for breadth and depth, and in developing ways of analysing large volumes of qualitative data, our imagining was for a new approach which, rather than re-inscribe methodological divisions, might bring them together.

1.5 Big Opportunities for Qualitative Research

So, let's first look at the areas where qualitative research might add value to big data, and examples of where this has been happening. We have shown that where big data analytics appear to struggle is in understanding the subjective and fine-grained nature and form of human relationships, interactions, and intimacies. If we wish to ensure that big data is collected and analysed in a meaningful way, then qualitative research surely has a rightful place and value. We concur with Grigoropoulou and Small (2022), who note that:

> Social science requires not only the ability to detect patterns in data but also the knowledge that the data are adequate and well understood, the patterns meaningful and appropriately interpreted, and the scientists themselves aware of their potential biases and limitations. If so, then the data revolution in social science will require, at its centre, the work of interviewers and ethnographers. (Grigoropoulou & Small, 2022, p. 3)

The interpretive skills of qualitative research have the capacity to make big data analytics richer and more nuanced (Housley et al., 2014, p. 12), and for Grigoropoulou and Small (2022) this can mean supporting all aspects of big data production, from understanding the human influences on why certain data was collected and others not, to how respondents choose to share their data, through to the ways these shape the data generated. Bail (2014), whose work examines cultural environments, has discussed the ways in which cultural sociologists can productively provide input into the development of computational models by assisting in the classification of cultural elements such as frames, symbolic boundaries, or cultural toolkits. Here, qualitative research is helping to make sense of the data, and in particular providing human understanding of the ambiguous patterns or 'noise' that can appear in data observations (Grigoropoulou & Small, 2022, p. 2). This input is especially effective in research teams that bring qualitative researchers and computational analysts into active collaboration (Mills, 2018), an issue we will talk more about shortly.

Further work is required to ensure the centrality of qualitative research in big data analytics, but there is a growing body of social scientific work exploring new big data sources and new techniques that demonstrate the intellectual possibilities. Much of this work is qualitatively driven and utilises new developments in computational software to 'dig' into latent meaning. Seale's work (Seale, 2006; Seale et al., 2006; Seale & Charteris-Black, 2008, 2010) on health and illness has been especially formative, not only in demonstrating the ways big samples can be qualitatively analysed using computer-assisted methods, but by sharing techniques that enable the analyst to move iteratively between quantitative and qualitative observations to gain fuller meaning.

Procter et al.'s (2013) 'real-time' analysis of Twitter during the 2011 London Riots is a further example of qualitatively driven big data studies. They analysed a data set of 2.6 million tweets using a combination of computational tools with established content analysis methods. This work was innovative in the way it used computer-assisted methods as the means through which to decide where to target human expertise in interpretation and analysis. Caton et al.'s (2015) exploration of German politicians on Facebook similarly used computational text analysis to examine their data qualitatively while at the same time providing 'scalability' not possible through manual methods. Others have engaged with social media to qualitatively explore digital communication and behaviour over time. Wilk et al. (2022) examined panic buying across the USA, UK, and Australia during the COVID-19 crisis. Here the computational package used, Leximancer (which we discuss in Chap. 5), effectively showcased how qualitative big data from social media can provide practical insights, using lexical analysis driven by machine-learning algorithms. Murthy's (2017) work combined quantitative and qualitative approaches to analyse Twitter data on a controversial song, with the qualitative component used to theoretically inform the manual coding of data.

This body of work showcases just a fraction of the research happening internationally intentionally using qualitative analytical frameworks and philosophies to make sense of large volumes of data. Together it shows the innovative potential for methods and methodologies that are not simply seeking to mix qualitative and quantitative approaches, but seeking to meaningfully integrate them. These studies are not concerned with methodological divides or conceptual grappling over what method should be prioritised. Rather, the primary concern is for designing a methodology best suited for handling the data in question and answering the research questions in the most meaningful way possible. The positive consequence is that it supports and enables a research community which is itself integrative, collaborative, and focused on joint goals.

1.6 In Search of Breadth and Depth

The examples in the previous section are working with big data, through qualitatively driven sociological investigation. In this sense they might be considered a different field of research from the breadth-and-depth method, since the latter is not an approach for working qualitatively with big data, but rather one that can enable you to gather, handle, and analyse large volumes of qualitative data. Nevertheless, understanding this field is crucial since it has helped form a path towards the breadth-and-depth method by showing the wealth and diversity of new potential data sources, and by illuminating the possibilities that come from combining big data with qualitative ways of thinking. There is some resonance here with Munk's (2019, pp. 159, 164) research on new Nordic food movements, where they used their research to explore the ways 'big and small' can work in mutual support of each other. They point to a 'curatorial approach' to analysis wherein the qualitative component adopts a "critically reflexive role" in reviewing and analysing the computational text analysis. In this way, computational tools and algorithms can be re-thought as "performing essentially qualitative work" whereby these mechanisms interact with the explorative, inductive work of the ethnographer. Burrell (2012), similarly, has developed a guide for ethnographers to understand and explore the possibilities of engaging productively with big data.

Another way of thinking about this field of work is that, rather than widening the methodological divide, it is productively blurring the boundaries of qualitative and quantitative research. Wiedemann (2013, 2017) is one such researcher whose work has specifically sought to highlight the methodological value of working qualitatively with large volumes of data. He notes that advances in the field of text mining are becoming increasingly sophisticated, and better able to extract meaning and context from text. The more this capacity develops, the more likely it will be that qualitative and computational text analysis can be integrated (Wiedemann, 2013). Computational analysis is thus not an 'end in itself', but rather part of a wider approach that meshes techniques and merges technical and methodological boundaries:

> It remains possible […] to use an initial exploration of the larger data sets which are now available as a way to pinpoint the specific areas that will prove most fruitful for 'small data' and 'deep data' research. (Bruns, 2013, p. 2)

This work introduces an integrative approach where qualitative and quantitative techniques are in active dialogue, prompting critical examination of what it means to get close to the data (Richards, 1998; Richards & Richards, 1994). While this wider integrative field has been developing (see, e.g., Andrade & Andersen, 2020; Lichtenstein & Rucks-Ahidiana, 2021) one of the concerns identified for the breadth-and-depth method was that while computational methods were useful for providing an overview of the data landscape, they often failed to access depth. Thick or close readings, if undertaken, felt like they were separate or static from the wider project. A methodologically integrative strategy, one that productively combined 'big' and 'thick' data, was not fully captured (Zhang et al., 2018). Brower and colleagues use the metaphor of binoculars to describe this, having the capacity to focus in or out to "closely examine phenomena at the micro or individual level and then dial out to view phenomena at the macro or societal level" (Brower et al., 2019, p. 5). This, they suggest, allows research to identify large-scale patterns and connect them meaningfully to analysis at the individual level. This absence in the literature was what moved us to recognise the importance of approaches that actively and meaningfully combined breadth and depth.

1.7 Big Qualitative Data Analysis

What we have discussed so far are ways in which qualitative research—its unique skills and analytical approach—can help add nuance and richness to the analysis of big data. We have also looked specifically at big data projects which adopt a qualitatively driven analytic strategy. Our own project emerged as a consequence of qualitative researchers interested in how best to go about analysing large volumes of qualitative data but doing it in a way that retains the distinctive characteristics of rigorous qualitative research. This question was becoming ever more relevant in the face of open science debates. Policy initiatives such as the *OECD Principles and Guidelines for Access to Research Data from Public Funding* (OECD, 2007) and the *Berlin Declaration on Open Access to Knowledge in the Sciences and Humanities* (Berlin Declaration, 2003) have set out requirements for the archiving of qualitative data from primary research projects for sharing and reuse, and promoted a drive towards data sharing more generally. Big data has, at the same time, provided qualitative research with new source materials and the computational tools to analyse large volumes of text data. These opportunities have prompted scholars to explore big qual and the opportunities that might come from large-scale qualitative data analysis.

Given its relative infancy, there is no agreed definition of precisely what constitutes big qual, and what volume of data is necessary for a project to fall under a big data categorisation. Brower et al. (2019) are amongst a small body of researchers explicitly defining their work as big qual, and advocate for the field to be fundamentally separated from big data. Their big qual study was a mixed methods project examining the delivery of developmental education in Florida and involved data from 542 research sites and 166 focus groups comprising 1100 participants. Given the size of the study, it involved a team of researchers to code the data qualitatively. They use a working definition of big qual based on a review of literature. Big qual, they suggest, is:

> data sets containing either secondary qualitative or primary data with at least 100 participants, analysed by teams of researchers, often funded by a government agency or private foundation, and conducted either as a standalone project or in conjunction with a large quantitative study. (Brower et al., 2019, p. 2)

This introduces a numerical element to the definition of big qual, which sits uneasily with a qualitative approach concerned with context and process. As noted earlier, our own working and non-prescriptive definition of big qual highlights process, that is, a project that involves working with a volume of qualitative data that is much larger than the quantity that would be feasible for a solo researcher or small team to collect and analyse themselves using the sorts of in-depth qualitative analytics approaches that we describe in Chap. 7. This was an issue explored by Bisel after it emerged in a conference panel at the 2013 National Communication Association annual meeting in Washington, DC. Together with colleagues, they hosted a round table event involving five scholars engaged in the big data field. Following extensive discussion about the social and cultural trends of big data and its implications for qualitative organisational communication research, the group concluded that the size of the data set was less of a concern in the definition than complexity and value:

> It seems that each of us—in our own way—is emphasizing the need to think carefully about the meaning of 'big' to include depth, richness, and theoretical importance. We are suggesting that the 'bigness' of data should be understood in relationship to the questions that can be answered and the theory that can be extended. (Bisel et al., 2014, p. 628)

What we have noted in our own review is that scholars discussing big qual are not only limited in number, but they are also associated with rapid analysis techniques aimed at expediting the analysis process, for example, rapid qualitative data analysis methods and matrix-based analysis (Taylor et al., 2018; Watkins, 2017). Musselwhite and Shergold's (2013) longitudinal study of driving cessation in later life is one such example tagged as big qual. Very large volumes of data were collected across the five waves and managed using a structured matrix mapping process. Similarly, Fontaine et al.'s (2020, p. 1) big qual study used "large qualitative data sets within a mixed methods study to add breadth, depth and context to an analysis". In their study, they developed a narrative approach to big qual by adapting the Listening Guide, a tool for tracing meaning in narratives, in the analysis of a set of 191 narrative text-based case notes. Here we have an emphasis on 'mixed methods' as a means of overcoming the limitations of either a purely large-scale quantitative or small-scale qualitative study.

Linabary et al. (2021) also used big *qualitative* data to refer to the quantity and complexity of their data. The study into activism related to domestic violence involved qualitative analyses of a large corpus of tweets in addition to interviews with women who participated in the hashtag #WhyIStayed. As they describe, analysis of tweets was 'paired' with interviews as a means of gaining both breadth and depth in their understanding of the hashtag and its possibilities and limitations as a form of feminist activism. This useful reflection on *complexity* takes us away from a definition of big qual that relates to a specific or set number or type of documents, or indeed specific methods. Rather, the definition relates the connection we can make between method and theory, and the capacity of the data set and methods to answer questions that would otherwise have been impossible to answer. This rather fuzzier definition is echoed in the work of Hossain and Scott-Villiers (2019), whose alternative perspective on big qual connects it to the rise of qualitative participatory research with large numbers of people across diverse settings. From this point of view, big qual relates not simply to size, but to the ability to access a greater number of research settings and contexts. This makes for a more complex form of social inquiry than conventional qualitative research.

These studies undoubtedly involve large volumes of qualitative data. Yet they continue to rely on what we might refer to as traditional approaches to qualitative data analysis which often comprise a full reading/listening of the data. In other words, conventional techniques are 'scaled up'. Such an approach becomes difficult—if not impossible—to apply to larger volumes of data. The issue is also analytical. Brower et al. (2019, p. 4) state that when working with such large volumes it becomes increasingly problematic to follow a cohesive narrative when the 'noise' drowns out the main 'story line'.

1.8 Computers, Computing, and Qualitative Data

In our big qual project, engagement in new computational processing tools was central. We saw these as introducing possibilities for interacting dynamically with the data and allowing the iterative movement between breadth and depth. Yet a common assumption is that qualitative research is somehow incompatible with engagement in computational analytics. But of course, qualitative researchers are not detached from technological developments, nor have they failed to engage in digital sources and resources. In fact, it is increasingly common for qualitative researchers to engage with large, digitised data sets, a tradition supported by the international expansion of archives (which we discuss in Chap. 3). Likewise, new technologies have supported qualitative researchers in developing rigorous procedures for data transcription, digitisation, and analysis using Computer Assisted Qualitative Data Analysis (CAQDAS).

Such procedures are exemplified by recent additions to CAQDAS software, such as automatic coding and sentiment analysis. We have also seen engagement in computer-assisted text analysis within the humanities and social sciences, although interaction with qualitative analysis remains limited (Wiedemann, 2017). Much CAQDAS has remained largely 'qualitative', however, with processes designed to replicate manual analytical procedures such as memoing, coding, and sorting data into categories (Wiedemann, 2013). In the meantime, committed qualitative researchers have continued to prioritise data quality over quantity, stressing qualitative research traditions as ever more imperative in the face of the big data:

> In a world where everything is measured, where sentiment and topics can be analysed on mass, and with increasingly sophisticated algorithms—providing a human context is crucial. (Conner, 2015, np)

Caution about the reliance on computers continues to feature in qualitative research teaching, and academic writing, and there is general agreement that such programs cannot be a substitute for the analytical and interpretative work of the human researcher. As already discussed, the most enduring arguments relate to the perceived mechanistic epistemological position of computing technology as being in opposition to qualitative researchers' objectives of seeing social phenomena from the perspective of human actors (Chowdhury, 2015). The perceived concern is that the volume and procedures associated with computer-assisted software means that it operates with analytical distance, which can result in context and meaning being lost:

> implicit assumptions of the software architecture will interfere with the qualitative research process and will result in the loss of shades of meaning and interpretation that qualitative data bring. (Rodik & Primorac, 2015, p. 1)

These concerns come together with discomfort associated with becoming 'computational subjects' (Berry, 2012) and the associated loss of the human or 'natural' aspects of the 'craft' of qualitative analysis, whether that be pens, paper, or close reading.

1.9 Capabilities and Skills

Mills (2019) has called on researchers working with qualitative research to participate in opportunities created by big data: to be 'open' to new computational techniques, and be 'ready' to think, discuss, and do our research differently. While we agree with this call, we also concur with Elliot, who, writing in 2005, pointed to the value to be gained from quantitative researchers being more 'open' to engaging in qualitative praxis and its philosophy. This, she suggests, could encourage a "more reflexive approach to the presentation of research findings by quantitative researchers" (Elliot, 2005, p. 186). There is clearly space for greater openness and collaboration on both sides of the methodological spectrum. The new forms of data becoming available, and the methods being developed, have the potential to support this change, and 'shatter' the disciplinary 'silos' that exist in the field of big data (Bail, 2014, p. 468). For King (2014) computer-assisted methods have the potential to end the quantitative-qualitative divide by bringing researchers together to work collaboratively on common problems. Such a development, King argues, can strengthen the role and impact of the social sciences.

Capacity and training does, however, remain an issue. Computational methods are complex methods, and those designed to extract latent meaning and sentiment require advanced training. There are limited tools and guidance available to support novice researchers navigating computational methods in big data projects. Felt

(2016), for example, found that many researchers wishing to engage in social media data analysis did not have the necessary skill set. This means that for those without a background or training in statistics or computer programming these methods can remain out of reach (Bail, 2014, p. 467). Wiedemann (2017) observed a lack of basic methodological and theoretical knowledge specifically with respect to qualitative research in this field. Qualitative researchers are not unique—many social scientists, regardless of methodological orientation, do not have these skills. Yet given the rise of big data and the possibilities it offers the social sciences, the value of text analytics is likely to grow—as will demand for its application.

Issues of skills and capacity also relate to qualitative research training. As Brannen (2017) notes, our research training, and subsequent careers, tends towards the development of either qualitative or quantitative skills. There are still relatively few mixed methods researchers, and while there is scant research on the topic, it is likely that most computational social scientists will have minimal training in qualitative data analysis, and therefore lack the interpretative and social theory skills necessary to inform their programming and interpretation of results.

A connected issue is that big qual is an emergent field, with partially developed methodological praxis. Linabary et al. (2021, p. 723) articulate this well in reflecting on their own experience as qualitative researchers on a big data project as being "fraught with onto-epistemological baggage and contradictions". Others have pointed to a scepticism amongst certain qualitative researchers towards quantitative research methods and vice versa (Albris et al., 2021). At times, our own involvement in a big data project puzzled both qualitative and quantitative researchers, a response which suggested a gap in methodological knowledge *and* disciplinary culture. To bridge this chasm, investment is needed in training, capacity building and multi-disciplinary working across the methodological spectrum. This would involve computational scientists and quantitative researchers gaining experience and expertise in qualitative approaches and sensibilities, and qualitative researchers developing skills in computational techniques.

A fruitful and generative space for skills and relationships to develop will be in interdisciplinary teams, projects, and research centres (see, e.g., Lichtenstein & Rucks-Ahidiana, 2021). Depending on the nature and size of the project, this could, for example, include incorporating skills in computational text analysis directly into bids for funding—either through a named collaborator equally invested in the project outcomes, or sufficient resources for the training and development of the team. Equally, this could include resources for training in qualitative research philosophies and techniques. An overall concern for collaborative projects is that those managing projects do not separate team members into methodological silos. Rather, research teams should be supported and enabled to invest all aspects of the project, thereby supporting connections to be made across analytical levels. Albris et al.'s (2021) approach was to ensure all types of information were shared across the team in a format that was simple and easy to understand.

While broadly we recognise and support the need for further training, the breadth-and-depth method can be conducted without in-depth training in computational methods. As we discuss in Chap. 6, there are various text-mining packages that all researchers can use. We introduce several that are accessible, easy to learn, and, importantly, free. Whatever the size of your project, the breadth-and-depth method can adapt.

As we will discuss in detail, the breadth-and-depth method demands engagement in both qualitative and computational approaches. It is notable, therefore, that so much attention is devoted to encouraging qualitative researchers to engage in computational techniques. The same question might also be asked of quantitative researchers in relation to their skills in qualitative research. This takes us into deeper questions around how, as researchers, we see and reflect on our own skills. From the perspective of the breadth-and-depth method, we recognise the importance of skills development across the methodological spectrum. Even more than this, we see big qual as providing opportunities for interdisciplinary and collaborative working.

1.10 Conclusion

In this chapter, we have looked at the place and value of qualitative approaches in working with big data. Within this field of interest, big qual is a space where qualitative researchers have the freedom to construct their own approach to defining, analysing, and theorising 'big' data. This is an emergent field, and those working within it are utilising a range of different approaches and data. We have sought to provide a broad definition of big qual by identifying its key features. Qualitative and quantitative components of the research are often discussed as doing different types of work—the qualitative element accesses 'rich' descriptions of micro processes, and the quantitative seeks patterns that explain macro processes. Such assumptions are not only incorrect, but they can separate qualitative and quantitative methods (Venturini & Latour, 2010). There is far more to research than skills and method, and researchers—regardless of method—can come together through their philosophy, theory, and reflexivity. In designing the breadth-and-depth method, we have been keen to develop those approaches that seek to meaningfully integrate these aspects of research. This can be addressed through interdisciplinary frameworks, and team-based research where different skills are valued. In the next chapter, we introduce the breadth-and-depth method, with the subsequent chapters covering each stage in detail.

1.11 Resources

This resource includes contributions from researchers working internationally who are engaged in the new big qual research field:

Weller, S., Davidson, E., Edwards, R., & Jamieson, L. (2019). *Analysing large volumes of complex qualitative data: Reflections from international experts*. NCRM Working Paper http://eprints.ncrm.ac.uk/4266/

This blog discusses how big qual can be deployed as a distinct research methodology to develop new forms of qualitative research and elucidate complex interactions between large-scale qualitative data sets:

Jamieson, L., & Lewthwaite, S. (2019). Big Qual—Why we should be thinking big about qualitative data for research, teaching and policy. *LSE Blogs*. Big Qual—Why we should be thinking big about qualitative data for research, teaching and policy | Impact of Social Sciences (lse.ac.uk)

References

Albris, K., Otto, E. I., Astrupgaard, S. L., Gregersen, E. M., Jørgensen, L. S., Jørgensen, O., Sandbye, C. R., & Schønning, S. (2021). A view from anthropology: Should anthropologists fear the data machines? *Big Data & Society, 8*.

Anderson, C. (2008, June 23). The end of theory: The data deluge makes the scientific method Obsolete [online]. *Science*. Retrieved December 12, 2022, from The End of Theory: The Data Deluge Makes the Scientific Method Obsolete | WIRED.

Andrade, S. B., & Andersen, D. (2020). Digital story grammar: A quantitative methodology for narrative analysis. *International Journal of Social Research Methodology, 23*, 405–421.

Andrejevic, M. (2014). Big data, big questions| the big data divide. *International Journal of Communication, 8*, 1673–1689.

Bail, C. A. (2014). The cultural environment: Measuring culture with big data. *Theory and Society, 43*, 465–482.

Baker, S., & Edwards, R. (2012). *How many qualitative interviews is enough*. Discussion Paper. National Centre for Research Methods. Retrieved December 12, 2022, from https://eprints.ncrm.ac.uk/id/eprint/2273/

Benjamin, R. (2019). Assessing risk, automating racism. *Science, 366*(6464), 421–422.

Berlin Declaration. (2003). *Berlin declaration on open access to knowledge in the sciences and humanities*. https://openaccess.mpg.de/Berlin-Declaration

Berry, D. (2012). Introduction: Understanding the digital humanities. In D. M. Berry (Ed.), *Understanding digital humanities*. Palgrave Macmillan.

Berry, D. M. (2011). The computational turn: Thinking about the digital humanities. *Culture Machine, 12*, 1–22.

Bisel, R. S., Barge, J. K., Dougherty, D. S., Lucas, K., & Tracy, S. J. (2014). A round-table discussion of "big" data in qualitative organizational communication research. *Management Communication Quarterly, 28*, 625–649.

boyd, D., & Crawford, K. (2011, September). Six provocations for big data, *A Decade in Internet Time: Symposium on the Dynamics of the Internet and Society*. Retrieved December 12, 2022, from https://doi.org/10.2139/ssrn.1926431

boyd, D., & Crawford, K. (2012). Critical questions for big data. *Information, Communication & Society, 15*, 662–679.

Brannen, J. (2017). *Mixing methods: Qualitative and quantitative research*. Routledge.

Braun, V., & Clarke, V. (2016). (mis)conceptualising themes, thematic analysis, and other problems with Fugard and Potts (2015) sample-size tool for thematic analysis. *International Journal of Social Research Methodology, 19*(6), 739–743.

Brower, R. L., Jones, T. B., Osborne-Lampkin, L. T., Hu, S., & Park-Gaghan, T. J. (2019). Big qual: Defining and debating qualitative inquiry for large data sets. *International Journal of Qualitative Methods, 18*, 1–10.

Bruns, A. (2013). Faster than the speed of print: Reconciling 'big data' social media analysis and academic scholarship. *First Monday, 18*(10).

Burrell, J. (2012). The ethnographers complete guide to big data: Small data people in a big data world. *Ethnography Matters*. Retrieved December 12, 2022, from http://ethnographymatters.net/blog/2012/05/28/small-data-people-in-a-bigdata-world/

Caton, S., Hall, M., & Weinhardt, C. (2015). How do politicians use Facebook? An applied social observatory. *Big Data & Society, 2*(2).

Chowdhury, M. (2015). Coding, sorting and sifting of qualitative data analysis: Debates and discussion. *Quality & Quantity, 49*(3), 1135–1143.

Conner, O. (2015). The history of qualitative research. *Medium*. Retrieved December 12, 2022, from https://oliconner.medium.com/the-history-of-qualitative-research-f6e07c58e439

Drucker, J. (2012). Humanistic theory and digital scholarship. In M. Gold (Ed.), *Debates in the digital humanities*. University of Minnesota Press.

Edwards, R., & Holland, J. (2013). *Qualitative interviewing*. Bloomsbury.

Elliot, J. (2005). *Using narrative in social research: Qualitative and quantitative approaches*. Sage.

Felt, M. (2016). Social media and the social sciences: How researchers employ big data analytics. *Big Data & Society, 3*(1).

Fontaine, C. M., Baker, A. C., Zaghloul, T. H., & Carlson, M. (2020). Clinical data mining with the listening guide: An approach to narrative big qual. *International Journal of Qualitative Methods, 19*.

Garcia, G. A., & Ramirez, J. J. (2021). Proposing a methodological borderland: Combining chicana feminist theory with transformative mixed methods research. *Journal of Mixed Methods Research, 15*, 240–260.

Grigoropoulou, N., & Small, M. (2022). The data revolution in social science needs qualitative research. *Nature Human Behaviour, 6*, 904–906.

Harford, T. (2014). Big data: A big mistake? *Significance, 11*, 14–19.

Hossain, N., & Scott-Villiers, P. (2019). Ethical and methodological issues in large qualitative participatory studies. *American Behavioral Scientist, 63*(5), 584–603.

Housley, W., Procter, R., Edwards, A., Burnap, P., Williams, M., Sloan, L., Rana, O., Morgan, J., Voss, A., & Greenhill, A. (2014). Big and broad social data and the sociological imagination: A collaborative response. *Big Data & Society, 1*(2), 1–15.

King, G. (2014). Restructuring the social sciences: Reflections from Harvard's Institute for Quantitative Social Science. *PS. Political Science & Politics, 47*(1), 165–172.

Kitchin, R. (2014). Big data, new epistemologies and paradigm shifts. *Big Data & Society, 1*(1).

Kitchin, R., & Mcardle, G. (2016). What makes big data, big data? Exploring the ontological characteristics of 26 datasets. *Big Data & Society, 3*(1).

Kracauer, S. (1952). The challenge of qualitative content analysis. *The Public Opinion Quarterly, 16*, 631–642.

Leurs, K., & Shepherd, T. (2017). Datafication and discrimination. In K. Van Es & M. T. Schäfer (Eds.), *The Datafied society*. Amsterdam University Press.

Lichtenstein, M., & Rucks-Ahidiana, Z. (2021). Contextual text coding: A mixed-methods approach for large-scale textual data. *Sociological Methods & Research, 0*(0).

Linabary, J. R., Corple, D. J., & Cooky, C. (2021). Of wine and whiteboards: Enacting feminist reflexivity in collaborative research. *Qualitative Research, 21*, 719–735.

Lupton, D. (2015). *Digital Sociology*. Routledge.

Marche, S. (2012). Literature is not data: against digital humanities. *Los Angeles Review of Books*. Retrieved December 12, 2022, from http://lareviewofbooks.org/article.php?id=1040&fulltext=1

Mason, J. (2012). How many qualitative interviews is enough. In S. Baker & R. Edwards (Eds.), *Discussion Paper*. NCRM. Retrieved December 12, 2022, from https://eprints.ncrm.ac.uk/id/eprint/2273/

Mayer-Schönberger, V. A. C. K. (2013). *Big data: A revolution that will transform how we live, work, and think*. Houghton Mifflin Harcourt.

Mills, K. A. (2018). What are the threats and potentials of big data for qualitative research? *Qualitative Research, 18*, 591–603.

Mills, K. A. (2019). *Big data for qualitative research*. Routledge.

Munk, A. K. (2019). Four styles of quali-quantitative analysis: Making sense of the new Nordic food movement on the web. *Nordicom Review, 40*, 159–176.

Murthy, D. (2017). The ontology of tweets : Mixed-method approaches to the study of twitter. In L. Sloan & A. Quan-Haase (Eds.), *The SAGE handbook of social media research* (pp. 559–572). SAGE.

Musselwhite, C. B. A., & Shergold, I. (2013). Examining the process of driving cessation in later life. *European Journal of Ageing, 10*, 89–100.
OECD. (2007). *OECD principles and guidelines for access to research data from public funding*. https://www.oecd.org/sti/inno/38500813.pdf
Playford, C., Gayle, V., Connelly, R., & Gray, A. J. G. (2016). Administrative social science data: The challenge of reproducible research. *Big Data and Society, 1;3(2)*, 1–13.
Procter, R., Vis, F., & Voss, A. (2013). Reading the riots on twitter: Methodological innovation for the analysis of big data. *International Journal of Social Research Methodology, 16*, 197–214.
Richards, L. (1998). Closeness to data: The changing goals of qualitative data handling. *Qualitative Health Research, 8*, 319–328.
Richards, T. J., & Richards, L. (1994). Using computers in qualitative research. In N. K. Denzin & Y. S. Lincoln (Eds.), *Handbook of qualitative research* (pp. 445–462). Sage.
Rieder, B., & Röhle, T. (2017). Digital methods: From challenges to Bildung. In M. Schäfer & K. Van Es (Eds.), *The Datafied society: Studying culture through data* (pp. 109–124). Amsterdam University Press.
Rodik, P., & Primorac, J. (2015). To use or not to use: Computer-assisted Qualitative Data Analysis Software usage among early-career Sociologists in Croatia. *Forum Qualitative Sozialforschung / Forum: Qualitative Social Research, 16*.
Seale, C. (2006). Gender accommodation in online cancer support groups. *Health, 10*(3), 345–360.
Seale, C., & Charteris-Black, J. (2008). The interaction of age and gender in illness narratives. *Ageing and Society, 28*(7), 1025–1045.
Seale, C., & Charteris-Black, J. (2010). Keyword analysis: A new tool for qualitative research. In I. Bourgeault, R. Dingwall, & D. Vries (Eds.), *The SAGE handbook of qualitative methods in health research*. Sage. Ch. 27.
Seale, C., Ziebland, S., & Charteris-Black, J. (2006). Gender, cancer experience and internet use: A comparative keyword analysis of interviews and online cancer support groups. *Social Science & Medicine, 62*(10), 2577–2590.
Silver, N. (2012). *The signal & the noise: Why so many predictions fail—But some don't*. The Penguin Press.
Smith, R. J. (2014). *Missed miracles and mystical connections: Qualitative research, digital social science and big data* (pp. 181–204). Emerald Group Publishing Limited.
Taylor, B., Henshall, C., Kenyon, S., Litchfield, I., & Greenfield, S. (2018). Can rapid approaches to qualitative analysis deliver timely, valid findings to clinical leaders? A mixed methods study comparing rapid and thematic analysis. *BMJ Open, 8*, e019993.
Van Es, K., Coombs, N., & Boeschoten, T. (2017). Towards a reflexive digital data analysis. In K. Van Es & M. T. Schäfer (Eds.), *The Datafied society*. Amsterdam University Press.
Venturini, T., & Latour, B. (2010). The social fabric: Digital footprints and quali-quantitative methods. *Festival For Digital Life and Creativity*, Proceedings of Future En Seine 2009, Cap Digital.
Wang, T. (2016). *Why big data needs thick data*. Retrieved December 12, 2022, from https://medium.com/ethnography-matters/why-big-data-needs-thick-data-b4b3e75e3d7 2022
Watkins, D. C. (2017). Rapid and rigorous qualitative data analysis: The "RADaR" technique for applied research. *International Journal of Qualitative Methods, 16*, 1609406917712131.
Wiedemann, G. (2013). Opening up to big data: Computer-Assisted Analysis of Textual Data in Social Sciences. *Forum Qualitative Sozialforschung / Forum: Qualitative Social Research, 14*.
Wiedemann, G. (2017). Computer-assisted text analysis beyond words. In: S. Weller, E. Davidson, R. Edwards, & L. Jamieson (Eds), *Southampton, National Centre for Research Methods*. Retrieved December 12, 2022, from Analysing large volumes of complex qualitative data.pdf (ncrm.ac.uk).
Wilk, V., Mat Roni, S., & Jie, F. (2022). Supply chain insights from social media users responses to panic buying during COVID-19: The herd mentality. *Asia Pacific Journal of Marketing and Logistics*, ahead-of-print.
Wyly, E. (2014). Automated (post)positivism. *Urban Geography, 35*, 669–690.
Zhang, S., Zhao, B., & Ventrella, J. (2018). Towards an archaeological-ethnographic approach to big data: Rethinking data veracity. *Ethnographic Praxis in Industry Conference Proceedings*, 62–85.

Introducing the Breadth-and-Depth Method

2

2.1 Introduction

This chapter provides an overview of the breadth-and-depth method. In so doing, it builds on the rationale for and value of large-scale qualitative data analysis that we outlined in Chap. 1, where we situated big qual analysis within wider debates about the breadth and availability of big data. Synthesising multiple qualitative data sets creates opportunities for asking new questions of existing data, and for making comparisons between material generated in different contexts. Working across data sets can also increase the diversity of sample populations and contexts, thereby enhancing the potential for theoretical generalisability about social processes (Davidson et al., 2019).

The breadth-and-depth method is a response to the increasing availability of archived qualitative data sets, along with the growing impetus for researchers to both share and reuse existing material. Reusing existing data sets is regarded as essential to accountability and transparency, and for enhancing public trust in research (Mauthner et al., 1998; Bishop, 2009; Mauthner, 2012; Mauthner & Parry, 2013; UKRI, 2016; Slavnic, 2017; Hughes & Tarrant, 2020; Weller, 2023). The method is also a response to the emerging interest in the value of big qual (see Chap. 1). Our goal was to develop a method that fused computational text mining techniques for exploring breadth with conventional forms of qualitative analysis that provide in-depth insights, so that new questions could be asked of existing, multiple, and/or pooled data sets. In essence, the breadth-and-depth method is an iterative process comprising four interconnected steps with each guided by the outcomes of the previous. Whilst the subsequent chapters focus, in detail, on the intricacies of each step, it is important to first gain a sense of the method in its entirety.

In this chapter, we will:

- discuss the rationale for and development of the breadth-and-depth method
- outline the metaphorical foundations
- offer an overview of the method, including an illustrative example

© The Author(s), under exclusive license to Springer Nature Switzerland AG 2023

S. Weller et al., *Big Qual*, https://doi.org/10.1007/978-3-031-36324-5_2

- consider the different relationships between theory and method and suggest how the four steps in our method might work in each
- suggest some key areas for consideration throughout the process
- provide details of relevant online resources

2.2 The Metaphorical Foundations of the Method

The breadth-and-depth method is founded on an archaeological metaphor. Metaphors are figures of speech, embedded in everyday language, and used to describe an object, phenomenon, or process in a manner that, if used appropriately, aids and gives depth to understanding (Carpenter, 2008; Lakoff & Johnson, 2003). Metaphors can also be used to build or convey theories and models. We found the use of metaphor, and in particular archaeological field investigations, useful in our attempts to capture big qual data as a landscape, something that can be mapped in its breadth and as containing features of interest that can be investigated in more detail. In other words, we can combine extensive coverage with intensive illumination (Davidson et al., 2019).

Our use of archaeology as a metaphor emerged from and built on the work of Seale and Charteris-Black (2010), who explored keyword patterns to gain an overview of their large corpus of qualitative data. They described the process as comprising:

> … an aerial view of a landscape, whose undulations and patterns of vegetation growth reflect the outline of ancient buildings, only possible to see from the air. At this point, the 'aerial archaeologist' descends to ground level and starts to dig. Once key passages are identified by keyword analysis, the researcher 'descends' to do more detailed analytic work, using procedures with which qualitative researchers are more familiar. (p. 537)

An archaeological metaphor enabled us to view a collection of big qual data sets as a landscape that needs to be scanned to gain an overview of the breadth of the material and as comprising interesting features worthy of more detailed examination. This metaphor also encouraged us to consider how we access data, at different levels and in alternative ways, and to think about what lies 'underneath' the corpus of material being analysed. It also helped us to work both extensively and intensively to identify and excavate meaning.

Figure 2.1 illustrates how the breadth-and-depth method is analogous to an archaeological field investigation. The method commences with processes akin to aerial reconnaissance with the researcher flying systematically across a data landscape. A map (i.e. a set of research projects) aids the researcher in gaining an overview of the textures and features of different land masses (i.e. projects). Once these are surveyed, the researcher then undertakes a ground-based, more detailed geophysical survey of the surface of the most promising area (i.e. selected collection of data sets) to assess what merits closer investigation. This is followed by the digging of shallow 'test pits' where it is thought, based on the previous surveys, artefacts (i.e. data of relevance to a research question/project) may be found. If a test pit does

Fig. 2.1 A four-step archaeological process moving between breadth and depth. (Illustration created by Chris Shipton https://www.chrisshipton.co.uk/)

not yield material of interest, then the researcher returns to the geophysical surveying to locate alternative areas of interest. The final task concerns archaeological deep excavation exploring in detail specific cases of interest, focusing on depth rather than breadth. In the breadth-and-depth method, work is conducted iteratively, and the process involves zooming out to the aerial view, and zooming in to the depth view, simultaneously.

2.3 An Overview of the Four Steps

A brief description of each step, the connections between steps, and the iterative nature of the process are outlined below. In addition, Fig. 2.2 provides an overview of the process, along with an illustrative example of a gendered and generational breadth-and-depth investigation of home moves we conducted using data sets from the Timescapes Qualitative Longitudinal Data Archive (for a detailed description of this work see Edwards et al., 2021a).

2.3.1 Step One: 'Aerial Surveying'—Overviewing the Qualitative Data and Constructing a Corpus

The first step in the breadth-and-depth method comprises an enquiry-led overview of relevant qualitative research. This could include data sets that have been digitised

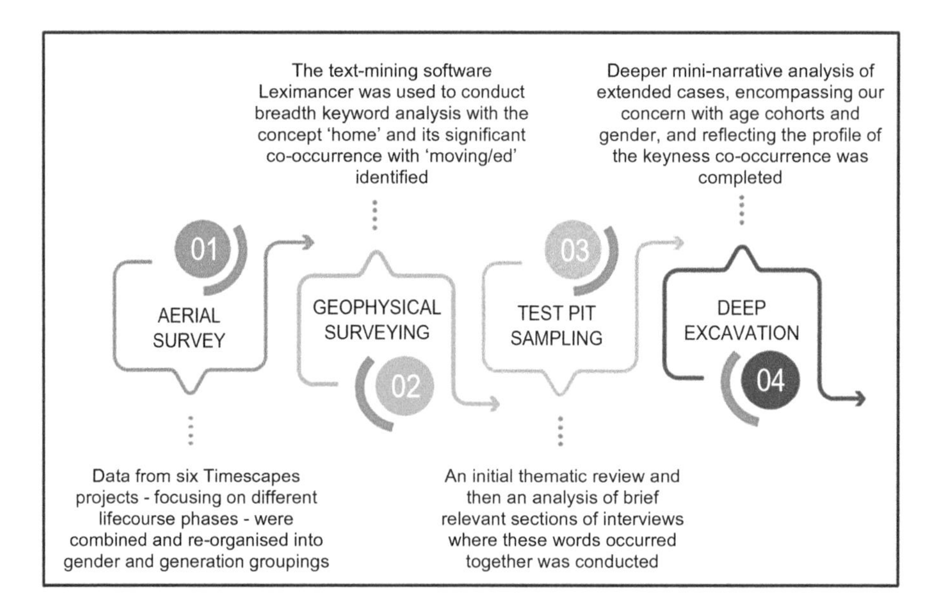

Fig. 2.2 Using the breadth-and-depth method to explore a data assemblage comprising material from the Timescapes Data Archive

and housed in a research repository or material pooled as part of a collaboration with other researchers. The step is concerned with sourcing and synthesising multiple qualitative data sets and/or organising a large volume of qualitative material in order to create a new data assemblage. This step can be likened to an archaeological aerial survey with the researcher flying systematically across a vast data landscape to gain a broad overview. Whilst it might be tempting to explore elements of the data in more detail, the purpose of this step is to survey the data landscape to identify material that might be pertinent to your research question(s). You may wish to survey data from a range of local or national repositories, bring together your own data sets with those of others, or work with colleagues to synthesise material from a multi-sited project. Examples of qualitative data archives containing digital data are outlined in Chap. 3. You could assemble data sets in their entirety or select cases of relevance, for instance by sampling using participant characteristics or by geographical location or by the method with which data were generated. Your research questions may require the selection of data that affords opportunities for relevant comparison between these characteristics, locations, or methods.

This step may involve the exploration of data sets that are deposited in a single or several repositories with which you are unfamiliar. It is important then to examine the archived project metanarratives of the original studies. Utilising metadata in this manner is akin to an archaeologist using photographs in an aerial survey. Project metanarratives include stored information about the aims of the original study, papers, and reports published by the original teams, and usually digitised too. These metanarratives may detail all or any of the rationales for the work and the methods employed, as well as explorations of the researchers' disciplinary background or epistemological stances. Some repositories curate such documents alongside the original data. Figure 2.3 is an example project metadata sheet currently used by a research group internal archive located at one of our institutions and is adapted from those commonly employed in archives such as the UK Data Archive (https://www.data-archive.ac.uk/). The metanarrative information will help you to determine whether the material is relevant for your proposed study.

It is also essential to explore the metadata of material in the data set including the data type and participant characteristics. This will enable you to determine whether data are of an appropriate nature, quality, and 'fit' with the research topic or question(s), and whether you wish to use a data set in its entirety or work with a subsample.

The next phase of step one concerns careful organisation and data management. Auditing the data by exploring and comprehensively recording the nature and type of data is essential to managing the substantial amount of material with which you are likely to be working. Software, including packages commonly used to aid qualitative analysis, may prove essential in managing and organising the digital data sets. It is also important to consider how and where you will store a large volume of material, especially if there are linked audio or visual files. Once identified, the collection of selected data sets constitutes the digital assemblage with which you work.

PROJECT METADATA

PROJECT DETAILS	
Title:	
Alternative title:	
Study reference:	
Access:	
Persistent identifier (DOI):	
Principal investigator(s):	
Researcher(s):	

SPONSORS AND CONTRIBUTORS	
Depositor(s):	
Sponsor(s):	
Grant number:	
Other contributors:	

TOPIC	
Topics:	
Keywords:	
Abstract:	

COVERAGE AND METHODOLOGY	
Time period:	
Dates of data collection:	
Country:	
Geography/area:	
Observation units:	
Population:	
Number of cases:	
Method of data collection:	
Sampling procedures:	
Kind of data:	

CITATION AND COPYRIGHT	
Citation:	
Copyright:	

Fig. 2.3 Example project metadata sheet

In Chap. 4 you can learn more about the process involved in step one including:

- different approaches to gaining a broad overview of a collection of data sets
- ways of synthesising data and determining what to include or exclude
- approaches to building, organising, and managing a new corpus

2.3.2 Step Two: 'Geophysical Surveying'—Approaches to Breadth Analysis Using 'Data Mining' Tools

Once an assemblage has been constructed, the next step comprises recursive surface thematic mapping using computer text analysis including so-called data mining. This constitutes the 'breadth' element of our breadth-and-depth method. The overriding aim is to identify potential areas of conceptual and substantive interest in order to assess what merits closer investigation. It can be likened to an archaeologist's ground-based geophysical survey, where features of interest that lie just below the surface of the data landscape are mapped. To achieve this, computational tools are used to investigate further these areas of interest in preparation for 'digging deeper' into the corpus in steps three and four. The growth in the availability of different computational tools designed to facilitate automated text analysis offers new approaches to aiding qualitative analysis. An example is using text mining to examine word patterns that may indicate themes in a set of text-based documents, whether treated as a single corpus or organised to make comparisons. Whilst this may seem at odds with many approaches to qualitative work, we argue that it can be a useful tool for exploring the breadth of a data landscape as long as the outcomes are not regarded as the endpoint, but rather as a means of identifying places to dig deeper into the data. Moreover, the use of the term 'mining' can be confusing as it implies deep investigations into the data. Rather, it is more appropriate to consider such techniques as enabling an extensive surface-level analysis, essential for gaining an overview but unsuitable for seeking intensive insights.

Chapter 5 details some of the types of software packages available to help provide a geophysical view of qualitative data sets (Kivunja, 2013). These include free-to-access text analytic tools, packages enabling semantic analysis, those concerned with conceptual or thematic mapping, and the use of programming languages such as R. Your choice is likely to be determined by your own skill set and the resources available. Despite differences in the mechanisms of each software tool, the majority employ algorithm-driven sifting and sorting of 'bags of words' in a set of texts in order to search for associations. Text mining tools process thousands of iterations to see how words 'travel together' through a text or set of texts. Such processes are premised on the notion that the terms surrounding a given word determine its meaning. Chapter 5 details a range of techniques such as word frequencies, concordance, co-location, proximity, and keyness that can be used.

The nature of the output from your text mining endeavours will depend on the software tools used. Software designed for text analysis can typically show keywords in context as in Fig. 2.4.

Other software produces two-dimensional visual maps of extracted material using word frequencies and co-occurrence to create clusters of terms, known as concepts, that are inclined to feature together in a text (see, e.g., Cretchley et al., 2010, Zaitseva et al., 2013). Figure 2.5 is an example of output from the text mining software Leximancer. The key concepts feature as words, which are then clustered into colour-coded themes. The 'hotter' colours (reds, oranges) denote the more salient themes, whilst the 'cooler' colours (blues, greens) highlight less prominent themes.

```
      [P1_W1_Alisha_INT1_clean_1990_F.txt, 69]          As long as there is | love | between them they don't need
      [P1_W1_Alisha_INT1_clean_1990_F.txt, 86]   from another family but still | love | each other. Animals.
     [P1_W1_Alisha_INT1_clean_1990_F.txt, 136]                    ... the same | love | that I felt for my
    [P1_W1_Alisha_INT1_clean_1990_F.txt, 3933]                , the Internet. I | love | English. Everybody else hates
    [P1_W1_Alisha_INT1_clean_1990_F.txt, 3942]              else hates English. I | love | poetry. I get As
    [P1_W1_Alisha_INT1_clean_1990_F.txt, 5438]                  said,' Oh I | love | you people so much'
    [P1_W1_Alisha_INT1_clean_1990_F.txt, 7765]            to move out because I | love | my family a lot.
[P1_W1_BrownAllie_INT1_clean_1990_F.txt, 1326]             yummy. We used to | love | that for pudding and my
     [P1_W1_Daisy_INT1_clean_1990_F.txt, 1715]          mobile saying,' I | love | you!' He doesn't
    [P1_W1_Jasmin_INT1_clean_1990_F.txt, 5990]          at my designing and I | love | art and I'm really good
       [P1_W1_Jay_INT1_clean_1990_F.txt, 5942]            a bit of comedy… | love | … horror… fighting.
       [P1_W1_Kate_INT1_clean_1990_F.txt, 414]                My mum. Cos I | love | her I suppose and she
      [P1_W1_Misha_INT1_clean_1990_F.txt, 466]                   ... I just | love | her. What would I
      [P1_W1_Misha_INT1_clean_1990_F.txt, 989]         a little while they all | love | you. Their house is
     [P1_W1_Misha_INT1_clean_1990_F.txt, 1558]           . It's beautiful. I | love | it. Yeah, she
     [P1_W1_Misha_INT1_clean_1990_F.txt, 2610]           somehow. That's why I | love | her so much, because
     [P1_W1_Misha_INT1_clean_1990_F.txt, 3322]              five years. So she'd | love | to go and see them
     [P1_W1_Misha_INT1_clean_1990_F.txt, 4332]                that. But what I | love | doing, being the youngest
     [P1_W1_Misha_INT1_clean_1990_F.txt, 4340]            being the youngest, I | love | getting loads of things,
     [P1_W1_Misha_INT1_clean_1990_F.txt, 4503]        pushing him downstairs. I'd | love | to do that. But
     [P1_W1_Misha_INT1_clean_1990_F.txt, 4557]                to me now. I'd | love | to have a younger sibling
     [P1_W1_Misha_INT1_clean_1990_F.txt, 6576]              kinds of pets. I'd | love | to live on my own
     [P1_W1_Misha_INT1_clean_1990_F.txt, 6684]                , well, she will | love | me inside, but not
       [P1_W1_Nas_INT1_clean_1990_F.txt, 4748]               , it's too much. | Love | the films and stuff but
 [P1_W1_ThomasAlannah_INT1_clean_1990_F.txt, 5]              No. Parents and | love | . I don't think you
[P1_W1_ThomasAlannah_INT1_clean_1990_F.txt, 54]          didn't care for me and | love | me, then they wouldn't
      [P1_W2_Anne_INT2_clean_1990_F.txt, 1052]                     .. No! I | love | to be the youngest.
   [P1_W2_Bethany_INT2_clean_1990_F.txt, 6388]                 Oh my God, I | love | # mum'..
     [P1_W2_Danielle_INT2_clean_1990_F.txt, 79]            always good. Yeah I | love | movies. Like the ones
  [P1_W2_Danielle_INT2_clean_1990_F.txt, 1009]                       ...' I | love | it!' Yeah.
  [P1_W2_Danielle_INT2_clean_1990_F.txt, 1271]                  and lots... | love | it! That's pretty much
  [P1_W2_Danielle_INT2_clean_1990_F.txt, 5158]               together... I | love | stars, they're excellent and
  [P1_W2_Danielle_INT2_clean_1990_F.txt, 9781]          Festival. AWSOME! I | love | it and I've been every
      [P1_W2_Holly_INT2_clean_1990_F.txt, 165]              a girl who is in | love | with a vampire. it's
```

Fig. 2.4 Example output from Antconc

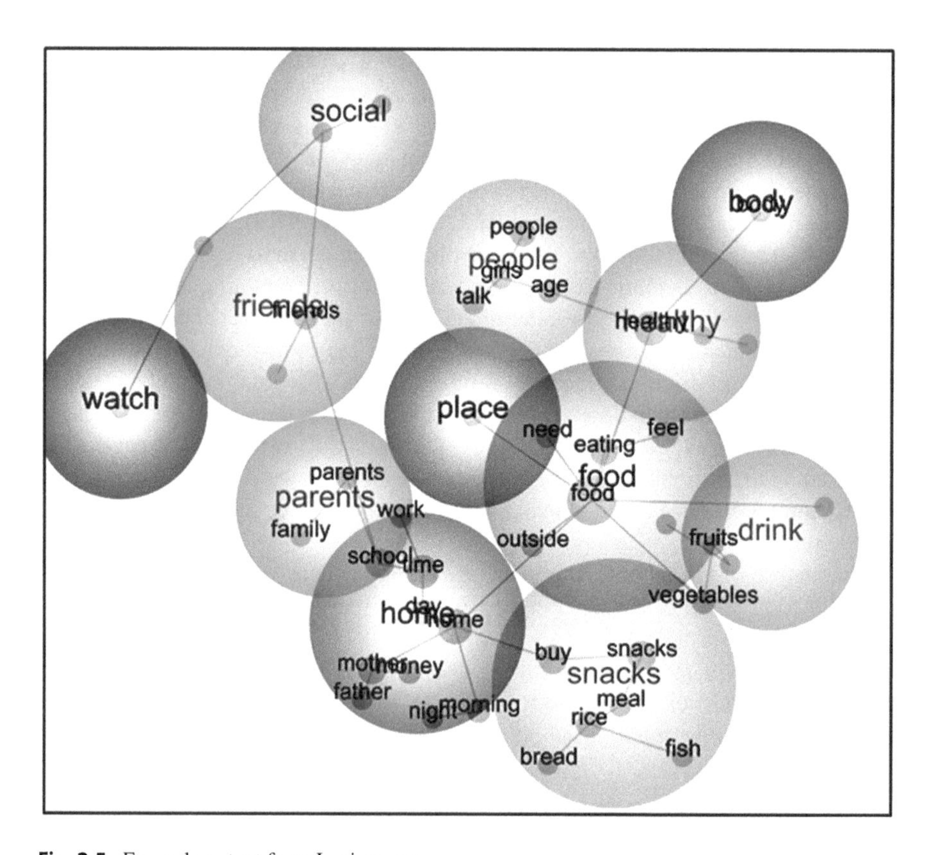

Fig. 2.5 Example output from Leximancer

Step two often involves working iteratively moving between the outputs and the original data to explore the context surrounding a keyword or concept. The outcome of this is the generation of an often large number of short data extracts. It is with these short snippets or brief glimpses into the data that step three is conducted. It is possible that you may not identify enough areas of conceptual and substantive interest, in which case it is advisable to return to step one to source further data sets of relevance.

Chapter 5 offers deeper insights into step two including:

- the processes, skills, and resources needed
- different approaches to text mining
- key concepts such as frequencies, concordance, co-location, proximity, and keyness
- examples of the types of software that can be used

2.3.3 Step Three: 'Test Pit Sampling'—Preliminary Analysis

After the recursive surface 'thematic' mapping of step two has been completed, the next step involves conducting an analysis of multiple small samples of likely data. In essence, step three is concerned with moving from breadth to depth. It comprises the sampling and preliminary analysis of short data extracts generated during step two. Drawing on our archaeological metaphor, this step is akin to digging a shallow 'test pit' deep enough into the data to judge whether or not there is material of relevance. Whilst it may be tempting to start to engage with the data in-depth at this point, the purpose of this step is to dig just deep enough to identify data of salience in preparation for more in-depth analysis in step four. You could, at this stage, be faced with a substantial number, perhaps hundreds, of extracts. These will need to be sampled for usefulness and salience to determine areas worthy of further investigation. Attention will need to be given to your sample size and this will be shaped by your sampling logic, which, in turn, is determined by your research design, as well as practical considerations of time and resources. Examples of approaches to sampling include theoretical, purposive, and realist. These complex issues will be addressed in Chap. 6.

This third step requires only a cursory reading of the short data extracts. This often involves exploring the text immediately surrounding the extract, only enough to gain a sense of the meaning and context in which the keyword or concept has been used. This will help determine whether the extract is of relevance to your research interests and questions. The length of the extract may be determined by the type of data with which you are working. For instance, in studying extracts from interviews with individuals we have tended to explore segments akin to a paragraph (around 150 to 200 words). Whilst working with data generated through focus groups the extracts tended to be shorter. Even at this shallow stage, the context in which extracts were generated remains central. Any keywords or concepts that seem ambiguous or have multiple meanings, or lack purchase for your research focus, should be eliminated from the analysis. Prior to moving on to step four, it may also be fruitful to sample keywords or concepts that seem less striking but that may be of potential value to your analysis. Equally, if a test pit does not result in any material of interest, then you will need to return to step two to conduct further mapping.

Chapter 6 provides detailed guidance on step three including:

- the process of digging deep enough into the data to see whether there is anything of interest
- key matters for consideration such as sample size and sampling logic
- undertaking cursory readings and eliminating ambiguous extracts
- approaches to recording and organising relevant extracts
- moving from breadth to depth (and back again)

2.3.4 Step Four: 'Deep Excavations'—In-Depth Interpretive Analysis

Step four is concerned with conducting in-depth analysis of the type familiar to qualitative researchers. With respect to our archaeological metaphor, this step may be likened to 'deep excavations' into key areas of salience identified during the preliminary analysis of 'test pits' conducted in step three. In this fourth step, the emphasis is on moving to greater depth through immersion in whole cases using conventional approaches to qualitative analysis. A case is the unit of analysis with which you work. Chapter 7 provides a detailed explanation of what constitutes a case. Importantly, this step involves a sensitivity to the context in which the data was generated, and to the detail and multi-layered complexity that constitutes qualitative work.

This step is concerned with examining in detail material from a range of textual and visual sources. This is likely to involve multiple readings or explorations. Depending on the approach employed, data may be reorganised and reordered, to explore, identify, and interpret patterns, meanings, and processes and the relationships between them. The approach you select will be dependent on your epistemological stance, the theoretical framework guiding your work, your substantive concerns, your research questions, along with practicalities concerning the configuration of the data. There are many options including, but not limited to, discourse, framework, narrative, and thematic analysis. You may wish to employ more than one approach to explore different facets of the material. During step four, other areas of relevance may come to light, and you may wish to return to step two to explore further the output from the recursive surface mapping.

Chapter 7 provides detailed insights into the potential approaches to step four including:

- what constitutes a case and how to select cases
- an overview of different approaches to analysis
- the process of selecting one or more approaches

- important aspects for consideration including epistemological stance, conceptual approach, substantive concerns, along with the practicalities of the form(s) of data

Each of the four steps is guided by the outcomes of the previous. Keywords or concepts generated using text mining tools in step two shape the preliminary analysis and examination of the short data extracts in step three. These data extracts are then used as the basis for the in-depth analysis conducted in step four.

2.4 The Relationship Between Theory and Method[1]

In contemplating how you might utilise the method in your own work, it is important to consider the relationship between theory and method (see also Blaikie, 2007; Timmermans & Tavory, 2012; Åsvoll, 2014; Kennedy & Thornberg, 2017). Theory is important in all research endeavours in providing a conceptual framework for posing, pursuing, and answering questions about how and why society and social situations work in the way that they do. It can be important for both qualitative and quantitative investigations in providing systematic thinking that moves research beyond the descriptive in making knowledge claims. A key advantage of the breadth-and-depth method is that it can be applied whatever the theoretical approach or topic.

In this section, we outline four logics of inquiry: deductive, inductive, abductive, and retroductive. Whilst different authors conceive of these logics in alternative ways, we provide a summary of the common features and describe how the four steps of the method might work in each (for a more detailed description see Edwards et al., 2021b).

2.4.1 Deduction

In the deductive model of the relationship between theory and data, researchers will start from an existing theory, and adopt it as an analytical lens to guide attention to detail in a specific substantive field. The logic of movement is from the general theory to the specific empirical: taking data, applying a theoretical framework to them, and then using that theory to deduce a 'why' explanation for the empirical findings. In a deductive approach to research utilising the breadth-and-depth method researchers can undertake step one—the identification of sets of archived data suitable for secondary analysis, driven by the issue of whether or not each discrete data set and its construction into a merged corpus enable the hypothesis or theory to be applied to the specific substantive focus of the project.

The mapping of keywords and identification for themes of step two, gaining a sense of the data corpus, would be guided by the theoretical framing. In a weak form

[1] Discussion in this section draws on Edwards, R., Davidson, E., Jamieson, L., and Weller, S. (2021) Theory and the breadth-and-depth method of analysing large amounts of qualitative data: a research note, *Quality & Quantity*, 55: 1275–1280.

of deductive reasoning, alertness to the relevance of particular words as 'keywords' or particular word co-location and clustering as 'themes' would be guided by the theoretical framework. In a strong form, searches would proceed using predetermined words, word clusters, and co-locations. The extracts of text for step three investigation would be those that show particular promise of shedding light on how the theory works in relation to the study's research questions, and the selection of materials for the in-depth analysis of step four similarly showing how theoretical 'why' processes play out in richer detail. A singularly minded deductive approach to the breadth-and-depth method is likely to work through the steps in order, as a guided and straightforward process, rather than needing to move backwards and forwards between them, with the aim of testing ideas about evidence. There remains the possibility, however, that by the end of step two researchers may realise that their chosen corpus cannot reveal anything relevant to their hypothesis and need to return to step one.

2.4.2 Induction

The logic of an inductive model is that the meaning interpreted from data is the basis for inferring theoretical statements about the nature of the social world and generalisation of substantive findings. The logic of movement is from the specific empirical to the general theory. Researchers are as open as possible to theory emerging from data without preconception about the outcomes. Thus, the step one precursory understanding of the nature and quality of the available sets of archived qualitative data will be shaped by an inductive logic that is fairly wide ranging, locating data on a broad topic area that fits with the secondary analysis research topic and the potential for generating theory. Step two mapping across the corpus would involve identifying frequent words and themes that are generated by the outcomes of text mining or automated semantic analysis without a framing of the parameters and nature of the search. The next step, three, would be to undertake preliminary examination of features of the recursive mapping that occur together frequently, following the regularities that offer the potential for inferring theory about the identified 'what' and 'how' issues in greater depth in step four. An inductive logic is also likely to involve a more iterative approach to the breadth-and-depth method as interesting features and patterns are identified in steps three and four. In turn, that may require a return to step one and identification of more, different, data to merge into the corpus for exploring processes in other contexts in an effort to develop stronger theoretical generalisations about how social processes unfold.

2.4.3 Abduction

Abduction looks for and explores potential 'what', 'why', and 'how' explanations, seeking or abducting theory through the identification of theory gaps and data anomalies. The abductive model involves an iterative filling of a theoretical gap in a particular substantive field, putting together theories from quite different fields, moving back and forth between data and theories, making comparisons and interpretations, and rethinking and refining best possible, plausible explanations. It is a logic of movement that actively seeks out and moves from general theory gap to specific empirical puzzle (in light of existing theories) to novel theory explanation. An abductive logic in step one of the breadth-and-depth method will seek out the potential of the available relevant archived data sets for revealing unusual phenomena that have the potential to fill a theoretical gap and to stimulate the abductive process. Step two is likely to involve bringing together searches based on lists and also open-ended searching for patterns and regularities. However, in the latter case, the regularities would enable the pointing up of where there were irregularities and unexpected coincidences to pursue. Such anomalies could be followed up with close attention to examination of data extracts in step three, to see if there were indications of plausible 'what' and 'why' theoretical explanations. Indeed, abduction is often applied in qualitative interview research through close attention to passages of text, with ideas about meaning and significance abducted from the segment. Cases for deeper examination in step four would be chosen because step three indicated that they held out the promise of insightful comparative differences and unusual features to enable the building of new theories or pulling together of two quite different theories to fill a gap in knowledge. Recursive moving between steps is a feature; for example, if provisional and depth examinations in steps three or four showed up a surprising juxtaposition of features, this would entail revisiting the keyword themes search of step two to look for relevant irregularities that would then point to places in the corpus to explore initially in provisional and selectively in greater depth, which in turn might suggest other anomalies to pursue.

2.4.4 Retroduction

Logics of inquiry are often idealised, sanitised versions of the way qualitative research proceeds. In reality, theory and data analytic processes may be quite messy, as we have indicated above. Deductive researchers often are open to rethinking and challenging theory, while inductive researchers are never a completely uninformed tabula rasa. Inductive logic can involve an element of deduction when working with the data, such as a prior orientation, or deducing further research questions to explore during analysis. Deductive logic can allow testing of conclusions at different stages of the research process, which in terms of the breadth-and-depth method theory may be identified through induction applied in steps one to four, and then steps two and three repeated deductively using that theory. Further, all three of the relationships between theory and data outlined above may be going on simultaneously. Abductive logic purposefully utilises unusual features of deductively or inductively generated findings to develop plausible explanations and generate new theory. In practice or as planned, there are also what are referred to as retroductive logics to the relationship between theory and data, which posit complementary or overarching combinations of deductive, inductive, and abductive in the oscillation, backtracking, and creative process that is social research (see various perspectives on this in Åsvoll, 2014; Chiasson, 2001; Kennedy & Thornberg, 2017; and Timmermans & Tavory, 2012). Whatever the logic—following pure deductive, inductive, or abductive approaches, or a purposeful or eclectic retroductive process—the iterative steps of the breadth-and-depth-method of analysing extensive amounts of qualitative data can encompass flexibly a range of articulations between theory and data.

2.5 Benefits of the Method for Qualitative and Quantitative Researchers

The breadth-and-depth method enables social researchers across the qualitative–quantitative spectrum to gain new insights from accumulating large volumes of qualitative data. To do this, the method combines computational techniques and analytical approaches typically labelled 'quantitative' with approaches that are unequivocally 'qualitative'. Researchers in both traditions must, therefore, leave their comfort zones and in the process learn new skills and perhaps also new ways of thinking. Although the method starts from and is designed for big qual, researchers who self-define as 'qualitative' may have to move beyond their comfort zone from step one if their approach to qualitative research creates hesitation about 'secondary analysis'. Step one requires acceptance that new questions can be fruitfully and ethically asked of data already gathered in others' qualitative projects without denying the original dialogues between researchers and research participants that shaped them. Qualitative researchers must also learn to make use of computational methods and for some this will require setting aside a distaste for techniques that start with a lack of awareness of the context of data collection.

An adjustment of mindset from the beginning may also be required of the researcher who self-defines as 'quantitative'. How radical this is will depend on their prior orientation to qualitative data and sympathy with the canon of qualitative research. For example, there is little point in bringing together multiple small-scale in-depth projects exploring fuzzy issues, aspects of subjective meaning making, and the entangled complexities of social processes, without believing in the intrinsic value of such work. Even if the learning journey does not deliver everything hoped for, the methodological insights gained may have lasting legacies, for example facilitating appreciation of and capacity to read across a wider range of research, broadening research imagination.

For qualitative and quantitative researchers, this method affords the opportunity to consider questions which can only be adequately answered with the depth of detail offered by small-scale qualitative research but without collecting new data. Rather the questions are answered through secondary analysis of large volumes of qualitative data such as an assemblage of data sets from multiple small projects with different research participants. Given the growing archives of qualitative research projects that have been carefully curated, it is increasingly likely that existing qualitative data will 'fit' a researcher's new research questions, even if these questions are not the focus of the original projects (Hammersley, 2010). The existing volume of archived qualitative research is likely to enable questions that address enduring concerns such as poverty, inequality, or racism, albeit the necessary critical mass of research currently is less certain for some pressing concerns, such as climate change. We have argued that reuse and secondary analysis that reassembles existing data can be as original and creative as producing primary data (Lewthwaite et al., 2023), and in this and the previous chapter we have drawn on work which effectively illustrated theoretical and substantive gains of bringing together qualitative data in a new assemblage drawn from different projects (Dodds et al., 2021; Edwards et al., 2021a; Davidson et al., 2019; Irwin et al., 2012; Purcell et al., 2020; Seale & Charteris-Black, 2008).

For qualitative and quantitative researchers, the possibility of addressing questions through comparative analysis is opened up by combining data across different projects. New data assemblages can select projects in a broadly similar topic area but with different profiles of research participants in order to ask comparative questions such as 'How do different groups talk about ....?' or 'How do people living in different localities experience ...?' Projects conducted with similar participants at different historical times or at different ages and stages of the life course might be used to ask comparative questions about change over time. Such questions might be seeking knowledge of practical value, for example for policy makers looking to customise messages or interventions, or be seeking evidence to confront theoretical assumptions within disciplinary debates about underlying processes of social change. Projects with similar topics and research participants, but different methodologies, might be assembled to contribute to addressing comparative questions about theory-method-evidence links. We reject an approach which treats secondary analysis as if it is second best and are not advocating secondary data analysis because prospects of gathering primary data are poor, although this may be a factor in favour of its choice for some researchers or for particular projects. The breadth-and-depth methodology is not necessarily any less demanding than the cycle from preparing for and gathering primary data, through analysis to writeup, as this depends on the scale and ambition of the specific project.

2.6 Key Considerations

2.6.1 Epistemological Issues

As we discussed in detail in Chap. 1, the breadth-and-depth method was developed, in part, as a response to the seismic shift in interest in the possibilities of big data analysis (Mills, 2018). Concurrently, these moves have resulted in ontological and epistemological shifts in ways of knowing (boyd & Crawford, 2012) and a concern that the growing emphasis on big data will embed further the privileging of positivist epistemologies (Glenna et al., 2019). We were concerned that if qualitative work was not represented adequately in this burgeoning field, it might simply be regarded as supplementary to big (quantitative) data or become sidelined (Glenna et al.,

2019). Constructivist and interpretivist approaches can and need to be a part of big data discussions (see also Burrows & Savage, 2014).

2.6.2 The Nature of Data in the Breadth-and-Depth Method

It is important to consider the nature of data analysed using the breadth-and-depth method. The process and tools lend themselves to qualitative textual data, which often comprises verbatim transcripts from in-depth interviews or focus groups. Diary data or that generated through other, more creative, means could also be used but, in essence, the focus is on working with audio data transcribed into text or the written word. A transcript for research purposes is a rendering into a written or typed, digital or paper version of material that is created from audio and/or video recordings of the collection or generation of data (for detailed discussions see Hepburn & Boden, 2017; Jenks, 2011). For the purposes of the breadth-and-depth method, this is largely concerned with material from interviews, where a researcher or researchers and an individual or a group of research participants are involved in verbal exchanges of questions and responses.

The extent of 'what and how' detail from and about the interview interaction that is represented in a transcript to form the data for a project is determined by theoretical and philosophical assumptions and decisions of the original researcher/s, so the secondary analyst using the breadth-and-depth method is dependent on these. Transcripts can never be complete; they are inherently partial, but there are levels of granularity: Is it verbal language exchanges only that have been transcribed for the original research project—in full or selectively? Is every um, er, and moment of silence represented? Are laughs, sobs, and sighs included? Are flat, rising, falling, joyful, sarcastic tones recorded? And are non-verbal aspects such as smiles, grimaces, and body movements noted? The brief extract from an interview transcript in Fig. 2.6 contains indications of words, emotions, pauses, and emphasis.

Nonetheless, there remains a subsequent layer of decision-making for the secondary researcher about what represents data from those primary transcriptions—the complete transcripts or only elements of them. This is a decision that needs to be considered in the planning stage of the project. The stepped process that we elaborate for the breadth-and-depth method relates primarily to the analysis of words, especially in step two of the process, though it may come back into play for step four. Even with access to transcriptions with detailed representation of the spoken

INTERVIEWER: So that's something new since we last spoke I'm guessing.
INTERVIEWEE: Yeah that's true [laughter]. Yeah ... that's the only thing on the agenda at the moment ... try and find a job, save, so that we can go and find somewhere to live.
INTERVIEWER: Do you think that will still be in the London area?
INTERVIEWEE: Umm I'd like to think it would be in the London areas but we're not TOO sure because it would be nicer to go a little bit out of London.

Fig. 2.6 Transcript extract

elements of an interview, there are decisions to be made about what to include for analysis. Notably, is the data for analysis the transcription of the words spoken by the interviewee only, or does it include the questions posed by the researcher? A major issue here is the nature of the relationship between the researcher and the social world they are seeking to understand, with debates between those who regard themselves as collecting data in interviews and focus on their content and those who regard data as generated in interviews and thus focus on the interview form.

The first conception views researchers as able to access direct objective knowledge about the interviewee and their social world through interviewing them. In challenge, the latter conception, known as the radical critique, argues that interviews are not a method of grasping the unmediated experiences of research participants—that is, the content of the interview data. It is the enactment of the method, of interviewer and interviewee exchanges, that is data—the form. Some see the focus on form as posing an unbridgeable divide between the experienced and the expressed and call for attention to content and the ways that interview data may be used to discuss the social world beyond the interview encounter (Hughes et al., 2020), while others urge reflexive analytic attention to interviews as speech events because there cannot be content without form (Whitaker & Atkinson, 2020). Where the breadth-and-depth method analyst is positioned in this debate philosophically will shape the decisions about whether, for example, the questions asked by the interviewer are included or excluded in the keyword and keyness analyses, as we note in Chap. 5.

2.6.3 Quality

The 'quality' element of qualitative research, what it is and how it may be recognized, is subject to ongoing and vigorous discussion, and to attempts to develop indicators for identifying it and strategies for managing it. Quality debates and practices in working with big qual engage with the breadth-and-depth method in two main ways: in terms of the robustness of the primary data that is being used for the secondary research, and in relation to the rigour of the analytic practice of using the breadth-and-depth method.

There are a range of types of arguments that promote and attempt to establish quality in qualitative research. These span across: (i) adapting and amending the sorts of robustness criteria used in quantitative research—reliability, validity, and generalization—so that they apply to qualitative research; (ii) developing quality standards of rigour that are grounded in qualitative research, for example transferability, credibility, and resonance, and can be applied to all forms of qualitative work; (iii) rejecting a generalist approach and arguing that the epistemological and methodological principles and assumptions underpinning specific qualitative methods are so distinct that they each require their own quality criteria, for example focus groups as against observation; (iv) positions that bridge the universal and specific, identifying some quality markers that can apply across the spectrum of qualitative methodologies but need to sit alongside more nuanced, method-based understandings; and (v) arguments against quality guidelines and indicators for qualitative research as constraining innovation (Flick, 2018; Lester & O'Reilly, 2021).

Within each of the first four positions, researchers, research funders, research archives, and so on have sought to put forward and endorse various sets of quality criteria, checklists, and standards that can be used to evaluate the robustness and rigour of qualitative work, usually in relation to primary research. There also have been attempts to set out rubrics by which to evaluate the quality and suitability of existing qualitative research data for secondary analysis (Sherif, 2018). For the secondary researcher working with data from primary research that has been peer-reviewed and funded, and deposited according to the requirements of archives, there is also the assurance that forms of quality check will have been built into these activities. Flick (2007, 2018) moves on from this standard preoccupation in thinking about quality in qualitative research, however. He argues that lists of criteria fragment quality into particular steps in the research process, such as sampling or analysis. Rather, Flick focuses on the research process as a whole, and on the promotion of quality rather than judgement of it. He proposes transparency and triangulation as key quality promotion strategies. These two procedures are apposite quality strategies for the breadth-and-depth method, both in terms of the various data being reused and the method of analysing big qual.

Transparency through primary researchers' documentation of the details of their data set and research process is an integral quality issue in identifying data for reuse; at the very least secondary researchers need to know about the parameters, generation, and depth of potential archived data sets in order to assess their rigour and relevance as part of step one of the breadth-and-depth method (Chap. 4) and it can aid them in pursuing good-quality ethical and careful secondary practice (Chap. 8). Further, transparency is a feature of the breadth-and-depth method as a stepped analytic process. The steps provide a transparent structure for the secondary analytic research practice, and throughout this book we provide guidance about analytic strategies and possibilities for each of them. Importantly for rigour, we also advise logging an account of these processes documenting a trail of information about decisions made in the breadth-and-depth steps, echoing the quality expectations built on transparency in identifying primary data for reuse.

Triangulation as Flick's other key proposal for quality management departs from the conventional constrained notion of a form of validation through different data points fixing onto an external reality. Rather, he conceives of triangulation as a quality management strategy that is systematic and comprehensive across the research process as a whole. This encompasses the levels of theoretical and conceptual perspectives, combining of methodological approaches and methods, and researcher understandings. Flick argues that triangulation can throw various lights on how an issue or phenomenon is constituted, extending knowledge through producing it at different levels, and thus contributing to quality in research. The breadth-and-depth method steps speak to this notion of triangulation. Working with big data can combine archived material from different primary researchers with different perspectives and methods, brought into engagement with the secondary researchers' own ideas. Whether a merged data set or single very large data set is worked upon, the analytic steps in the breadth-and-depth method each enable different aspects of the data to emerge—from the keyword mapping of step two, through the themes of step three, to the in-depth case analysis of step four, enabling the production of quality through the amalgamation of the whole process.

2.6.4 Generalisability

We have suggested that drawing on data across projects may create new possibilities for theoretical generalisation by facilitating understanding of how social processes are manifest in different contexts. The idea of 'theoretical generalisation' may not be familiar to all researchers. One of the challenges of a method which draws on techniques and approaches across the range of quantitative and qualitative is finding a common language. We recognise that 'representative' and 'generalisation' in the qualitative sense are very different to their use in quantitative practice. Projects gathering qualitative data from small numbers of people cannot reasonably claim that their sample represents any larger population in a statistical sense and hence cannot generalise to any larger population through claiming statistical representativeness. However, a small in-depth study might, nevertheless, suggest a theoretical relevance that enables some generalisation beyond the immediate participants. Mason (1996, pp. 153-159) noted that such claims of 'theoretical generalisation' cover a range of logics. They include tentative suggestions that findings concerning processes and meanings may apply in similar settings elsewhere or when people have similar circumstances and characteristics. Sometimes qualitative researchers make stronger claims, such as suggesting that their small study is a particular case which is ideally suited to demonstrate a key theoretical issue and hence the findings are of more general relevance (Mason, 1996, pp. 153–159).

The breadth-and-depth method can be used to explore and subsequently modify or strengthen any such claims. This would start from the first step of purposefully assembling a new data set. It might, for example, involve assembling a range of projects on the same substantive topic but with data gathered from participants with different characteristics or circumstances or in contrasting settings. It might involve seeking studies that offer theoretically contrasting or alternative cases to a study making a strong claim of particular interest. Depending on the theoretical issue at stake, time, place, and characteristics of samples may again be key issues or the method of gathering data might be a key source of variation within the assembled data. In addition to extending the possibilities for generalisation through building appropriately faceted data sets, the combination of computer analysis of breadth and a conventional qualitative research approach to depth potentially strengthens reliability, validity, and generalisability. This claim is made by Nelson (2020) (see Case Study 5.2), who, as in the breadth-and-depth method, advocates for an approach combining the use of computer programs to discover key 'topics' that then guide the researcher to sites for deeper qualitative analysis. While not sharing the emphasis Nelson places on 'bias', we accept much of what she says about the advantages:

> Because this step [a conventional qualitative analysis involving 'deep reading'] is both guided and backed up by numbers [...] both the researcher and the reader can be more confident in the particular interpretation developed by the researcher. This makes the process more efficient, but it also ensures the researcher will not skip over important passages because of fatigue or bias. Conversely, because this step involves a human actually reading the text, the numbers are given context and interpreted in a meaningful, more traditional sociological and theory-informed fashion. (Nelson, 2020, p. 26)

2.7 Getting Started

As you get started there are several areas worthy of contemplation. The following questions may help guide your thinking through the process.

Planning Your Analysis

- What do you hope to gain by bringing together different data files or data sets?
- From where will you source relevant data? Will you explore material archived in local or national repositories, or bring together your own data sets with those of others?
- What kind of data do you think might be appropriate? Will geographical context or sample characteristics be important? What of the epistemological stance or theoretical perspective of the original researchers?

Exploring Available Data Sets

- Do you have access to project documentation? What contextual information might you require?
- What do you know of the original research aims and rationale? What might be the implications of any differences in the foci of the original studies?
- What do you know of the metanarratives of the data sets? Does each file have accompanying metadata?
- What do you know about the sample(s)?
- How was data generated and how might you assess the quality of the material?

Working with Your Data Assemblage

- How will you determine which data/data set(s) to include?
- What new questions do you plan to ask of the data?
- Will your initial criteria for exploring the data be broad-ranging or narrow in focus?
- What logic of inquiry will you employ?
- How will you structure your data assemblage?
- What computational tools will you employ to undertake text mining/semantic analysis? Do you require any training, resources, or support?

- What sampling strategy will you employ for your preliminary analysis?
- What approach(es) will you take to in-depth analysis?

Managing Your Data Assemblage

- What do you need to consider in terms of the physical, network, and computer/file security and the storing and organisation of a large volume of material?
- Will you use a database or qualitative analysis software package to manage the data?
- If you are utilising archived data, how will you ensure you comply with the end user licence? What is permitted in terms of data storage and retention?
- What do you know of the original consent for reuse process? A clear strategy for file labelling and the auditing and documenting of files is also recommended.
- How and when will you acknowledge the investments made by the original researchers in creating the data set(s)?

2.8 Resources

The following papers provide further details about the breadth-and-depth method:

Davidson, E., Edwards, R. Jamieson, L., & Weller, S. (2019). Big data, qualitative style: A breadth-and-depth method for working with large amounts of secondary qualitative data. *Quality & Quantity*, *53*(1), 363–376.

Edwards, R., Davidson, E., Jamieson, L., & Weller, S. (2021). Theory and the breadth-and-depth method of analysing large amounts of qualitative data: A research note, *Quality & Quantity*, *55*, 1275–1280.

Edwards, R., Weller, S., Jamieson, L., & Davidson, E. (2020). Search strategies: Analytic searching across multiple data sets and with combined sources, In K. Hughes & A. Tarrant (Eds.), *Qualitative secondary analysis* (pp. 79–100). Sage.

Weller, S., Davidson, E., Edwards, R., & Jamieson, L. (2019). *Analysing large volumes of complex qualitative data: Reflections from international experts*. NCRM Working Paper. http://eprints.ncrm.ac.uk/4266/

The following podcasts, created in collaboration with the UK's ESRC National Centre for Research Methods, provide an overview of the breadth-and-depth method:

Digging deep! The archaeological metaphor helping researchers get into big qual https://www.ncrm.ac.uk/resources/podcasts/mp3/NCRM_podcast_Weller2.mp3; https://www.ncrm.ac.uk/resources/podcasts/?id=space-for-big-qual

This film was recorded at the ESRC National Centre for Research Methods event on 'Approaches to Analysing Qualitative Data: Archaeology as a Metaphor for Method'. 18 October 2016, Foundling Museum, London, UK.

Davidson, E., & Weller, S. (2016). *A layered archaeological approach to analysis across multiple sets of qualitative longitudinal data*. https://www.youtube.com/watch?reload=9&v=SXI638JhHGQ&feature=youtu.be

This guide was created to assist researchers through the process of managing qualitative longitudinal data.

Neale, B., & Hughes, K. (2020). *Data management planning: A practical guide for qualitative longitudinal researchers*. https://timescapes-archive.leeds.ac.uk/wp-content/uploads/sites/47/2020/12/Data-Management-Planning-2020.pdf

This data set was created as a teaching resource for the breadth-and-depth method. It comprises the data assemblage we constructed to help develop the method and is available for use in both teaching and research by registered users of the archive.

Weller, S., Davidson, E., Edwards, R., & Jamieson, L. (2019). *Big qual analysis: Teaching data set*. Timescapes Data Archive. https://doi.org/10.23635/14

To support the use of the data set, we have also created a comprehensive selection of resources for both students and teachers.

Lewthwaite, S., Jamieson, L., Weller, S., Edwards, R., & Nind, M. (2019). Teaching how to analyse large volumes of secondary qualitative data. *NCRM Online Learning Resource*. https://www.ncrm.ac.uk/resources/online/all/?id=20727

References

Åsvoll, H. (2014). Abduction, deduction and induction: Can these concepts be used for an understanding of methodological processes in interpretive case studies? *International Journal of Qualitative Studies in Education, 27*(3), 289–307.

Bishop, L. (2009). Ethical sharing and reuse of qualitative data. *Australian Journal of Social Issues, 44*, 255–272.

Blaikie, N. (2007). *Approaches to social enquiry: Advancing knowledge* (2nd ed.). Cambridge.

boyd, d., & Crawford, K. (2012). Critical questions for big data. *Information, Communication & Society, 15*(5), 662–679.

Burrows, R., & Savage, M. (2014). After the crisis? Big data and the methodological challenges of empirical sociology. *Big Data & Society, 1*(1).

Carpenter, J. (2008). Metaphors in qualitative research: Shedding light or casting shadows? *Research in Nursing and Health, 31*(3), 274–282.

Chiasson, P. (2001). Abduction as an aspect of retroduction. In M. Bergman, & J. Queiroz (Eds.), *The commens encyclopaedia: The digital encyclopaedia of peirce studies*. Retrieved December 12, 2022, from https://www.commens.org/encyclopedia/article/chiasson-phyllis-abduction-aspect-retroduction

Cretchley, J., Rooney, D., & Gallois, C. (2010). Mapping a 40-year history with Leximancer: Themes and concepts in the journal of cross-cultural psychology. *Journal of Cross-Cultural Psychology, 41*(3), 318–328.

Davidson, E., Edwards, R., Jamieson, L., & Weller, S. (2019). Big data, qualitative style: A breadth-and-depth method for working with large amounts of secondary qualitative data. *Quality & Quantity, 53*(1), 363–376.

Dodds, C., Keogh, P., Bourne, A., McDaid, L., Squire, C., Weatherburn, P., & Young, I. (2021). The long and winding road: Archiving and re-using qualitative data from 12 research projects spanning 16 years. *Sociological Research Online, 26*(2), 269–287.

Edwards, R., Davidson, E., Jamieson, L., & Weller, S. (2021b). Theory and the breadth-and-depth method of analysing large amounts of qualitative data: A research note. *Quality & Quantity, 55*, 1275–1280.

Edwards, R., Weller, S., Davidson, E., & Jamieson, L. (2021a). Small stories of home moves: A gendered and generational breadth-and-depth investigation. *Sociological Research Online, 0*(0).

Flick, U. (2007). *Managing quality in qualitative research* (1st ed.). Sage.

Flick, U. (2018). *Managing quality in qualitative research* (2nd ed.). Sage.

Glenna, L., Hesse, A., Hinrichs, C., Chiles, R., & Sachs, C. (2019). Qualitative research ethics in the big data era. *The American Behavioral Scientist, 63*(5), 560–583.

Hammersley, M. (2010). Can we re-use qualitative data via secondary analysis? *Sociological Research Online, 15*(1), Article 5.

Hepburn, A., & Boden, G. B. (2017). *Transcribing for social research*. Sage.

Hughes, J., Hughes, K., Sykes, G., & Wright, K. (2020). Beyond performative talk: Critical observations on the radical critique of reading interview data. *International Journal of Social Research Methodology, 23*(5), 547–563.

Hughes, K., & Tarrant, A. (2020). *Qualitative secondary analysis*. Sage.

Irwin, S., Bornat, J., & Winterton, M. (2012). Timescapes secondary analysis: Comparison, context and working across data sets. *Qualitative Research, 12*, 66–80.

Jenks, C. J. (2011). *Transcribing talk and interaction: Issues in the representation of communication data*. John Benjamins.

Kennedy, B. L., & Thornberg, R. (2017). Deduction, induction and abduction. In U. Flick (Ed.), *The SAGE handbook of qualitative data collection* (pp. 49–64). Sage.

Kivunja, C. (2013). Qualitative data mining and knowledge discovery using Leximancer digital software. *Lecture Notes on Information Theory, 1*(1), 53–55.

Lakoff, G., & Johnson, M. (2003). *Metaphors we live by*. The Chicago Press.

Lester, J. N., & O'Reilly, M. (2021). Introduction to special issue quality in qualitative approaches: Celebrating heterogeneity. *Qualitative Research in Psychology, 18*(3), 295–304.

Lewthwaite, S., Jamieson, L., Davidson, E., Edwards, R., Nind, M., & Weller, S. (2023, in press) Enhancing the teaching of qualitative methods: teaching the 'breadth and depth method' for analysis of 'big qual'. In M. Nind (Ed.), *Handbook of teaching and learning social research methods*. Edward Elgar.

Mason, J. (1996). *Qualitative researching*. Sage.

Mauthner, N. (2012). Are research data a common resource? *Feminists@Law, 2*(2), 1–22.

Mauthner, N., & Parry, O. (2013). Open access digital data sharing: Policies, principles and practices. *Social Epistemology, 27*(1), 47–67.

Mauthner, N., Parry, O., & Backett-Milburn, K. (1998). The data are out there, or are they? Implications for archiving and revisiting qualitative data. *Sociology, 32*, 733–745.

Mills, K. A. (2018). What are the threats and potentials of big data for qualitative research? *Qualitative Research, 18*(6), 591–603.

Nelson, L. K. (2020). Computational grounded theory: A methodological framework. *Sociological Methods & Research, 49*(1), 3–42.

Purcell, C., Maxwell, K., Bloomer, F., Rowlands, S., & Hoggart, L. (2020). Toward normalising abortion: Findings from a qualitative secondary analysis study. *Culture, Health & Sexuality, 22*(12), 1349–1364.

Seale, C., & Charteris-Black, J. (2008). The interaction of age and gender in illness narratives. *Ageing and Society, 28*(7), 1025–1045.

Seale, C., & Charteris-Black, J. (2010). Keyword analysis: A new tool for qualitative research. In I. Bourgeault, R. Dingwall, & D. Vries (Eds.), *The SAGE handbook qualitative methods in health research*. Sage. Ch. 27.

Sherif, V. (2018). Evaluating preexisting qualitative research data for secondary analysis. *Forum: Qualitative Social Research, 19*(2). Retrieved December 12, 2022, from https://www.qualitative-research.net/index.php/fqs/article/download/2821/4211?inline=1

Slavnic, Z. (2017). Research and data-sharing policy in Sweden—Neoliberal courses, forces and discourses. *Prometheus, 35*(4), 249–266.

Timmermans, S., & Tavory, I. (2012). Theory construction in qualitative research: From grounded theory to abductive analysis. *Sociological Theory, 30*(3), 167–186.

UKRI. (2016). *Concordat on open research data.* Retrieved December 12, 2022, from https://www.ukri.org/files/legacy/documents/concordatonopenresearchdata-pdf/

Weller, S. (2023). Fostering habits of care: Reframing qualitative data sharing policies and practices. *Qualitative Research, 23*(4), 1022–1041.

Whitaker, E. M., & Atkinson, P. (2020). Response to Hughes, Hughes, Sykes and Wright. *International Journal of Social Research Methodology, 23*(6), 757–758.

Zaitseva, E., Milsom, C., & Stewart, M. (2013). Connecting the dots: Using concept maps for interpreting student satisfaction. *Quality in Higher Education, 19*, 225–247.

Part II

An Enquiry-Led Overview of Qualitative Research Data Sets

Sourcing and Searching for Suitable Data Sets

3

3.1 Starting your Breadth-and-Depth Project

In Chap. 2 we provided an overview of the breadth-and-depth method and its different stages. As described, the starting point is an enquiry-led overview of potential sources of qualitative research. Multiple data sets are surveyed and synthesised to find appropriate qualitative material for your project. The endpoint is the creation of a new data assemblage upon which you can conduct further analysis. Of course, not only does the possibility of creating a new data set depend on the existing data, but the material you wish to use must also be identifiable, accessible, and, crucially, available to you for ethical reuse.

Thus, availability is the issue that this chapter will attend to, by focusing on initial preliminary strategies for identifying possible data sets of interest. In Chap. 1 we spoke about formalised repositories for data. These have developed to form an infrastructure for big qual development, so to that end we devote much of the discussion here to locating and searching archival sources. We are, nonetheless, keenly aware of the multiple forms of data that exist outside and alongside formal large-scale repositories. Such sources might include data stored in community or personal archives, or data typically associated with quantitative endeavours, such as social media or open questions in large-scale surveys. As part of your breadth-and-depth analysis you might also choose to access different forms and permutations of data, such as combining archived data with data from your own project(s); amalgamating two or more data sets from an existing multi-site project; or pooling data collected separately but connected by a substantive topic or theme. Together, archives and alternative approaches are creatively challenging the boundaries of qualitative data, and the prospects for its engagement in big data. We aim to open up these exciting possibilities here.

In this chapter we:

- begin by discussing what we mean by qualitative data and the different forms it can take within a big qual project

© The Author(s), under exclusive license to Springer Nature Switzerland AG 2023

U. Weller et al., *Big Qual*, https://doi.org/10.1007/978-3-031-[illegible]

- examine the development of qualitative data reuse, and in particular developments in research data archives in the UK, and internationally
- provide information on key research data archives, and how to access them
- consider alternative sources of data, and the range of opportunities that these might bring to your research
- discuss the parameters for identifying possible archives for use in your research

3.2 What Do We Mean by Qualitative Data?

As noted, you may already have a data set or combination of data sets that you plan to utilise in your breadth-and-depth project. If not, a helpful place to start is to think about the range of qualitative data sources available. When we ask 'What is qualitative data?' we most likely think about materials such as interviews, focus groups, field notes, or audio and video recordings produced for research projects. This is what might be understood as data that has been created or elicited by the actions of social researchers (Heaton, 2008). While not obviously considered 'qualitative', we would suggest that this definition should extend to self-reported materials, such as surveys, where open questions can be used to gain qualitative insight into behaviours, perceptions, and experiences. This is also the type of data that you will commonly find in formal social research archives, typically in the form of data produced as or transcribed into text and available to access in person, or via digital download. The English Longitudinal Study of Ageing, for example, records life history interviews alongside panel data. Similarly, the international Young Lives study has grown from a household Panel Survey to encompass qualitative longitudinal research, qualitative sub-studies, and school-based surveys (see Case Study 4.3 for more details). There is a wealth of other forms of qualitative research available to explore, including data which is 'found' or collected with minimal structuring by researchers (Heaton, 2008). Here, forms of data are as far ranging as letters, diaries, public documents, and other forms of correspondence, through to electronic media generated through webscraping and Twitter feeds (amongst others).

Table 3.1 illustrates the fantastic wealth of possible qualitative data sources. These will only grow in scale and accessibility as new digital technologies improve processes of digitisation and data architecture becomes more sophisticated.

3.3 Where Can I Find Data?

3.3.1 Introducing the Archive

We turn now to thinking about how to identify and access possible data sets. If you don't have your own qualitative data set(s) that you wish to work with, then you may begin your search in an archive. An archive is, at its simplest, "a repository of some kind" (Moore et al., 2017, p. 1). We might immediately think about formal archives, for instance those attached to libraries, universities, or research

Table 3.1 Types of qualitative data

• Data from structured, semi-structured, or unstructured interviews; focus groups; oral histories (audio/video recordings; transcripts; notes/summaries; questionnaires/interview schedules and protocols)
• Field notes (including from participant observation or ethnography)
• Unmoderated conversations and focus groups (video, audio, or transcripts)
• Maps/satellite imagery/geographic data
• Official/public documents, files, reports (diplomatic, public policy, propaganda, etc.)
• Minutes of meetings
• Government statistics
• Correspondence, memoranda, communiqués, queries, complaints
• Parliamentary/legislative proceedings
• Testimony in public hearings
• Speeches, press conferences
• Military records
• Court records; legal documents (charts, wills, contracts)
• Chronicles, autobiographies, memoirs, travel logs, diaries
• Brochures, posters, flyers
• Press releases, newsletters, annual reports
• Records, papers, directories
• Internal memos, reports, meeting minutes
• Position/advocacy papers, mission statements
• Party platforms
• Personal documents (letters, personal diaries, correspondence, personal papers)
• Maps, diagrams, drawings
• Radio broadcasts (audio or transcripts)
• TV programmes (video or transcripts)
• Print media (magazine and newspaper articles)
• Electronic media
• Published collections of documents, gazetteers, yearbooks, etc.
• Books, articles, dissertations, working papers
• Photographs
• Ephemera; popular culture visual or audio materials (printed cloth, art, music /songs, etc.)
• Data generated from webscraping (i.e. from Twitter data and other social media, online forums, and discussion groups)
• Marginalia (text in the margins of documents or surveys)

Adapted from Types of Qualitative Data (syr.edu)

institutions, established for record keeping and the preservation of data. These spaces are typically governed by legislation relating to data privacy and confidentiality, bureaucratic procedures, and institutional codes of practice. In the UK, for example, the national archives are governed by the Data Protection Act (2018) and General Data Protection Regulation (GDPR). Measures such as these can benefit archives by recognising the public interest in permitting the permanent preservation of personal data for the long-term benefit of society, and importantly establish legal safeguards relative to how data is processed, stored, and made available for reuse. In such contexts, formal archives hold particular importance for the reuse of qualitative data since they afford researchers access to data which has already been subject to various checks and balances relative to ethical data handling, data privacy, and safeguarding.

While formal archives, and legislation and institutional codes that govern them, can provide reassurances about the data being accessed, archives have multiple characters and states of being. As Moore et al. (2017) notes, archives are not simply sites with many different types and forms of documents. They are also formed by and hold ideas, discourses, memories, and heritage (Moore et al., 2017, p. 1). These are not neutral or fixed collections of data, but rather reflect the wider political, cultural, and societal settings in which they were conceived and constructed. In their cross-country review, Besek et al. (2008) provide examples of archives that have encountered censorship, or the unnecessary removal of material considered to be a national security threat, controversial, offensive, or politically sensitive. In Case Study 3.1, an example from Burundi, we see processes of data collection aligning to political ideologies, and concerns that this relates to processes of control, rather than the regulation aimed at improving accessibility.

These examples remind us that we must give careful consideration to the wider administrative, social, and political contexts of the archive when accessing research of interest. These will have a bearing not only on the type of data preserved, but also its nature and form. It is also worth reiterating that the context in which an archive exists may be diverse and that this need not be a formal or institutional space. Rather the archive can be many things, variously a "building, cardboard-box, photograph album, internet website" (Moore et al., 2017, p. 1). Such alternative archival sources present unique opportunities for big qual analysis, and we consider these later in the chapter.

3.3.2 Qualitative Archives—A Growing Infrastructure

Those new to qualitative research might be surprised to learn that while there are long-standing institutional archives, prior to the mid-1990s accessible qualitative archives and re-use of qualitative data was limited. Archives tended to preserve quantitative data, or be the site of public and historical records, such as Census data. Data elicited by human researchers (e.g. interviews, focus groups, field notes, observations, questionnaires, and solicited diaries) was more typically the creation of individual researchers or research teams, and on completion either retained in private files or destroyed.

A 1991 study by the Economic and Social Research Council (ESRC) (the UK's largest funder of economic, social, behavourial, and human data science) showed that as much as 90 per cent of qualitative research data was not being formally archived. Data that was retained often failed to meet basic archiving standards related to accessibility, security, and cataloguing (Thompson, 1991). Meanwhile, writing in the USA in 2010, Elman and colleagues noted that while university libraries and research institutes were archiving research data, there was no infrastructure or agreed set of standards or practices for archiving or reuse. As a result, very few qualitative studies conducted by scholars in the USA were used more than once, with re-use almost entirely concentrated in quantitative studies (Elman et al., 2010, p. 23). Similarly, Mozersky and colleagues (2020b) note that qualitative health data in the USA is still rarely, if ever, shared. This will be changed as a consequence of new requirements set out by the National Institutes of Health (NIH). These state that data—regardless of whether qualitative or quantitative—should be made freely available as widely as possible. While this is a positive first move towards encouraging reuse, Mozersky et al. (2020a, 2020b) found that many researchers remain concerned about the lack of guidance, infrastructure, and resources required to deal with qualitative data.

Mozersky et al.'s (2020b) research reflects wider barriers to qualitative data archiving and reuse. Despite advances in training on approaches to archiving as well as in qualitative data analysis more generally, knowledge and access on reuse has not been promoted internationally, and until relatively recently there has been little appetite amongst researchers. While in certain cases, inertia in the preservation and reuse of qualitative data may relate to methodological culture, it has also been a consequence of prohibitive legislation relating to copyright, ownership, data protection, duties relating to confidentiality, and possible financial and resource implications. For many researchers these complexities have tended to outweigh the possible benefits of qualitative archiving.

This state of affairs sits in contrast with the quantitative research paradigm where secondary analysis is an established tradition and analytic practice sees numeric data as a publishable asset and an 'investment to the future' (Corti & Thompson, 2004, p. 3). Those seeking to advance qualitative data archiving argued that *not* preserving qualitative data was a lost investment, both financially and intellectually. Such losses were important for qualitative researchers to address, since quantitative data was unproblematically considered a better 'return' on research funding. This was a clear departure from previous thinking, which emphasised data destruction over data preservation.

In the UK, the creation of 'Qualidata' was a critical moment for qualitative archiving and reuse. Established in 1994 by the ESRC, it was among the first dedicated national data services for storing and accessing qualitative data. At the same time, it established procedures for ethical data preparation, storage, and dissemination (Corti & Thompson, 2004). Qualidata not only advanced the qualitative data landscape in the UK but began the shift across disciplines in attitude and culture around qualitative secondary analysis. A further change in policy direction came in 1995 as the ESRC required data from *all* research projects that it funded, both quantitative and qualitative, to be deposited with the UK Data Service and made available for reuse. Researchers applying for funding had to demonstrate that any proposed research could not be conducted without first using existing archived data sets. Since then, Qualidata has been integrated into the UK Data Archive (UKDA) and renamed Qualibank, a further step towards recognising both the importance of qualitative archiving investment and the possible returns. Other countries have looked, and continue to look, to Qualidata as exemplifying good practice in qualitative archiving (Corti, 2000).

The activity in archive generation in the UK has come alongside methodological advancement. In Case Study 3.3 we introduce the Timescapes Project whose data we utilised to construct our own big qual corpus for analysis. As noted in the Preface, it was an important programme of work in the UK since it signalled the ESRC's commitment to preserving qualitative data and demonstrating and encouraging its reuse. The ESRC have since built on these foundations through the Secondary Data Analysis Initiative, a programme specifically aimed at delivering high-quality research by using existing data resources. A range of qualitative data sets, including the Timescapes Data Archive, has been promoted through this programme of work.

Outside the UK, advances were also finding ground. As well as the establishment of the Finnish Social Science Data Archive (Kuula, 2000) and the QualiService in Germany (discussed in Case Study 3.2), Bishop and Kuula-Luumi (2017) helpfully summarise key international advancements in qualitative archiving. These include several new, specialist qualitative collections, as well as established archives that are commencing programmes for qualitative data preservation. One of the key initiatives in Europe is the Consortium of European Social Sciences Data Archives (CESSDA). At the time of writing, the consortium is working with 21 member countries' data services and repositories with the aim of providing a quality research infrastructure for high-quality social science research. Critically, CESSDA has supported countries where qualitative data material has rarely been archived. The European Commission have also been central in encouraging access to, and the reuse of, digital research data generated by Horizon 2020 projects through the Open Research Data Pilot (ORD Pilot) following FAIR data principles (i.e. all research data should be Findable, Accessible, Interoperable, and Reusable).

Within the USA, there is no single infrastructure governing qualitative research data archiving, although the Inter-university Consortium for Political and Social Research (ICPSR) has been influential. Collecting data since 2011, it partners with several federal statistical agencies and foundations to create collections organized around specific topics. This is complemented by the Henry A. Murray Research Archive, an endowed, permanent repository which holds key longitudinal qualitative and mixed methods studies, and the Qualitative Data Repository at Syracuse, a dedicated archive for storing and sharing digital data collected through qualitative and multi-method research in the social sciences. Beyond the UK, Europe and the USA, the development of data archives is sparser, with programmes remaining focused on the deposit and preservation of quantitative data. Such examples include the Indian Council of Social Science Research Data Service, the Australian Data Archive (or ADA) discussed below, and the Social Science Japan Data Archive (SSJDA) at the University of Tokyo. Similarly, the Israel Social Sciences Data Center focused on quantitative data—although this closed in 2019 due to lack of funding.

3.3.3 Where Can I Find Archived Qualitative Data Sets?

In this section we move on to discuss the key qualitative data archives available. This is not intended to be a comprehensive list, but rather illustrate the current range and depth of data available. Our own project was conceived with the Timescapes Data Archive (see Case Study 3.3) in mind since it provided the volume and type of data desired to develop the methodological component of our project, and offered the possibility of pursing our substantive interest in gendered care and intimacy across the life course. You may be in a similar position and have already been alerted to data sets relevant to your research through conversations with colleagues, reading published work on secondary analyses, or via external sources, such as the media. Alternatively, you may be approaching your project with only a broad research topic or concept in mind, and you may wish to browse these ideas using a more exploratory or investigative approach.

If you have not yet identified a suitable archive, you might choose to begin by browsing the range of stores detailed below. One starting point is the Registry of Research Data Repositories (www.re3data.org), a global catalogue of repositories which covers data from a range of academic disciplines. Funded by the German Research Foundation and launched in 2012, the site aims to 'promote a culture of sharing, increased access and better visibility of research data'. Within Re3data you can browse and filter your search for appropriate repositories by subject, country, and content type, although the ability to filter by qualitative data is limited. A further point to note is that while Re3data provides information and direct links to possible repositories for your project, it cannot guarantee heterogeneity across or within the collections you access; nor can it ensure the quality of the data you might wish to use. It should also be noted that while Re3data is defined as a global registry, links to repositories outside Europe are limited. This limitation reflects how, and where, qualitative archiving has developed, and the unequal level of investment it has received in different parts of the world.

An alternative starting point is the UK Data Archive, which although UK based collects data relating to research conducted globally. This is the largest collection of social, economic, and population data, and at the point of writing, there were over 1200 qualitative and mixed methods studies across many forms of data, including in-depth interviews, diaries, anthropological field notes, visual materials, and audio recordings. There is also a wealth of responses to open survey questions, a resource which has increasingly become a site of investigation for qualitative researchers (see, e.g., Elsesser & Lever, 2011). While the UK Data Archive is unique in terms of the volume of the data stored, it has also helped develop and apply quality standards for data management and preparing data for deposit which is well organised, documented, preserved, and accessible. While there will be differences in format and quality of data over time, data preserved with the UK Data Archive is expected to conform to rigorous archiving standards and data protection laws.

There are several other institutional archives in the UK worth mentioning, although more comprehensive lists are available at the UK Data Archive (see also Hughes & Tarrant, 2019). Typically, such archives have developed from particular programmes of research, such as the Timescapes Data Archive as discussed in Case Study 3.3. Other archives of note include The National Social Policy and Social Change Archive stored at the University of Essex. Set up and funded by Qualidata and the Joseph Rowntree Foundation, it houses the raw data arising from contemporary social policy research and includes classic studies by researchers such as Peter Townsend and Elizabeth Bott. Mass Observation is another well-known collection.

Housed at the University of Sussex, data submitted electronically is available for viewing online through Mass Obs Online (to which libraries/universities can subscribe). It provides unique data and materials on everyday life in Britain dating back to 1937. A more recent archive example is the Health Experiences Research Group (HERG) Archive at the University of Oxford. The Group collects and analyses video and audio-recorded interviews with patients, carers, and other family members on their experiences of illness. Its collection currently totals around 3000 interviews, and covers over 80 different conditions and topics, and has already been used extensively as a resource for extrapolating evidence to contribute to health policy formulation (see, e.g., Ziebland & Hunt, 2014).

We can also report encouraging examples of archive advancement in Europe. These include the Danish Data Archive (DDA), the Finnish Social Science Data Archive (FSD), the Northern Ireland Qualitative Archive (NIQA), the Irish Qualitative Data Archive, the Norwegian Centre for Research Data, the Czech Sociological Data Archive (SDA), and the Swedish National Data Service. Using international archives can raise further challenges of searching for terms across different or multiple languages and then making appropriate translations. The investment and support given to archiving also varies internationally, as does the quality and range of qualitative data available. The beQuali Archive in France, for example, was inspired by the British Qualidata model, but has struggled to sustain itself, with criticism over the low usage and high costs (Duchesne & Brugidou, 2016). The GESIS Data Archive for the Social Sciences in Cologne, Germany, has similarly taken many years of feasibility studies to become established (Kluge & Opitz, 2000). The Austrian Data Archive, WISDOM (The Wiener Institute for Social Science Data Documentation and Methods), began to include qualitative data sets in 2007, and Smioski (2011) has written about the challenging funding environment in which this extension has taken place.

The case studies in this chapter showcase a selection of these developments. In 3.2 we have an account of the development of the QualiService in Germany. The infrastructure for this archive has gradually developed from the work of a research centre. Boosted by funding from the German Research Foundation, and the commitment of key scholars, the archive is continuing to expand and promote data sharing across the country. Case Study 3.4 meanwhile discusses the Irish Qualitative Data Archive (IQDA). Hosted by Maynooth University Social Sciences Institute, it has become a highly respected and central access point for qualitative social science data generated in or about Ireland.

Looking internationally, there are a number of well-known data collections relevant to those looking to combine multiple data sets, or work with large volumes of qualitative data. The Murray Research Archive, USA, for example, is a multi-disciplinary research data archive focusing on the study of lives over time. With over 380 data sets and 7000 datafiles currently, this national repository focuses specifically on data that illuminates women's lives and issues of concern to women. Most of the studies archived include in-depth interviews or at the very least some open-ended survey questions. There is also a priority for archiving data that has not been subject to extensive analysis, and for data sets that contain qualitative or interview data, or which are longitudinal in design (James & Sørensen, 2000). The Qualitative Data Repository at Syracuse University in the USA holds data from around 113 qualitative and mixed methods research projects at the time of writing and aims to address the lack of a data sharing culture and infrastructure gap.

In Australia, the Australian Data Archive (ADA) at the Australian National University is made up of several sub-archives, although qualitative research preservation remains underdeveloped. A recent Australian Research Data Commons-funded project, Transforming Research Communities, led by McLeod et al. (2020), explored approaches required to build and sustain community ownership of a new archive, as well as best practice principles for preserving and reusing qualitative data. In South Africa, the GALA Queer Archive (Gay and Lesbian Memory in Action) was established to preserve local LGBTIQ narratives, both public and private. As well as photos, letters, diaries, and other artefacts, GALA also stores research data. Due to capacity issues, the data is not digitised and must be viewed in person. These examples, however, show that developments are taking place internationally in archiving and making available existing qualitative data.

We can also take a wider look across archival collections. Universities are one site to individually acquire collections. These can include the work of past scholars and include correspondence and individual private archives (Doorn & Tjalsma, 2007). In some cases, these documents are being preserved and made available for reuse. One such example is the British National Cataloguing Unit for the Archives of Contemporary Scientists (NCUACS), in Bath, UK. Established in 1987, its aim is to locate, sort, index, and catalogue the manuscript papers of distinguished contemporary British scientists and engineers. The NCUACS closed in 2009, with work being continued by the Centre for Scientific Archives (CSA). There are many scholar-specific collections held by universities such as The Olive Schreiner Letters online, led by the Universities of Edinburgh and Leeds Beckett. The papers of academics may also be archived and available through the institutions in which they lectured, such as Tom Burns at the University of Edinburgh, and Michael Young at the Churchill Archive Cambridge. Accessing this type of specialist resource can be incredibly rich but can depend on the researcher having local knowledge about the material and archive, or require preliminary desk research into possible sources.

3.4 Community Archiving

We noted at the start of this chapter that it is important to think about 'the archive' in its multiple forms. While we applaud the work of formal archives associated with university libraries and research institutions, we want to recognise the equal importance of supporting smaller endeavours. Cook (2013) talks about four paradigm shifts in archiving—the most recent shift relating to the fact that data is now *everywhere*. As they note, we are all collating our own data, and it is therefore possible for us to reclaim our own data and, in turn, gain control of the knowledge typically held by archives and archivists. Vardigan and Whiteman (2007) have examined the question of who uses and gains access to the archive, noting that, with support, there is potential for archives and archiving to be expanded to a much wider group. Simionica (2018) meanwhile has looked at the concept of community archiving expanding globally alongside greater attention within archives to record the narratives and lived experience of marginalised groups and embrace ideas of decolonisation and social justice.

Community archives, such as GALA mentioned earlier, are part of this shift and are creating spaces in which communities and social groups can document and record their own history. Examples include indigenous archiving in Australia (Evans, 2018) and the recording of LGBT histories in North West England (Wright et al., 2021). Community archives are not simply repositories focused on storage, but rather the 'creation of living knowledge' where the ongoing process of creating the archive becomes as important as the data ultimately recorded and stored (Popple et al., 2020, p. 14). This shift disrupts the idea of the archivist being responsible for the processing, description, and preservation of data. Indeed, it creates disquiet in our venture since it emphasises archiving, and by implication reuse, as a participatory project. In our breadth-and-depth project we made steps to contact the original research teams. We consulted, for example, on views about the data being combined and re-used and sought to understand more about how they felt about sharing their data. We also asked for support to fill in the gaps we had in our own contextual knowledge, for example, in instances where data was incorrectly recorded. While this connection and dialogue with the original researchers was fruitful, it was not possible to extend this to the participants themselves. While privacy laws would make such a dialogue impossible in the context of our data, it is worthwhile reflecting on the possibilities that such conversations might bring to our research and the associated ethical implications. We spend time in Chap. 8 discussing these issues in detail.

Discussions about data quality, privacy, reuse of data, use of commercial platforms, and the centralisation of data impact on the development of community archives in the same way as collections in formal archives (Popple et al., 2020, p. 14). It is often the case that community archiving projects involve a partnership with a professional archivist, or support from academics. Key recent texts on community archiving say little, or nothing, about reuse, emphasising instead the important role that the archive plays in and for communities, and the possibilities for activism or local democratisation around the process of archiving (Popple et al., 2020; Bastain & Flynn, 2018). We do not wish to promote the idea of community archives being under-exploited resources, waiting quietly for researchers to 'discover' potential reuse opportunities. Rather, we see the community archive as a space where the breadth-and-depth method might evolve further to encompass a more collaborative and participatory approach. One such example is the Women, Risk and AIDS project, discussed in Case Study 3.5. Here, the research team discuss an innovative collaboration between academics, archivists, and activists interested in young women's sexual health and empowerment. Working with a set of interviews collected as part of a social research study conducted in Manchester, UK, in 1988–1990, the team sought to reunite the data with the communities that generated them years after. This was a process of engagement in 'careful ethics' designed to ensure data and outputs were shared collaboratively.

3.5 Problematising Archiving and Reuse

We have provided an overview of qualitative archive development and highlighted the complexity of the archive as a site of research. It is worth pausing at this moment to review the scholarly debates that have taken place relative to the challenges and rewards of qualitative data archiving. Archives, as we have shown, are constantly evolving both in terms of their infrastructure and the nature of the content stored. The examples provided demonstrate the ability of the archives to take us across time and place, through different disciplines and methodological approaches. We can also see that the form that qualitative 'data' takes is no longer limited to what might be considered 'conventional' sources, such as transcripts from individual interviews or focus groups. Rather, new collections are extending to promote the inclusion of diverse materials, ranging from field notes, diaries, and photos, through to audio recordings, videos, and blogs. The preservation of data in a range of forms makes the venture of big qual analysis an especially exciting, if not challenging, endeavour.

The researchers' relationship to archiving and reuse is not straightforward. The case in support of reuse has been centred around gaining greater transparency in the research process and data analysis, heightened accountability, and improved dialogue and collaboration across research topics (Chauvette et al., 2019). Others have focused on the importance of bringing qualitative research in line with quantitative research, a field which continues to dominate the material archived. Reuse, from a purely pragmatic view, would allow the significant financial and temporal investment made in qualitative research to be realised (Hughes & Tarrant, 2019, p. 4). These benefits have been subject to contestation, with debates at the other side arguing that qualitative data sets cannot—from an epistemological, methodological, and ethical perspective—be reused. This points to the distinctive nature of qualitative research and the relative closeness that researchers develop with their participants and the data generated. This not only means that they may be more reticent about sharing their data than quantitative peers (Fink, 2000), but that the corporeal experience of 'being there' gives the primary research team an understanding of the data that secondary analysts can never possibly have. But it is also the case that qualitative researchers may feel more exposed, and their skills and processes under examination. Thus Elmand et al. (2010, p. 24) note that while transparency provided by archiving and reuse can "encourage scholars to be more self-conscious and rigorous in how they undertake data collection", it might also have the "unintended consequence of encouraging researchers to engage in self-censorship, omitting data collected using a technique they fear will not be considered rigorous, or failing to record relevant observations that they worry may reflect badly upon themselves or a subject".

While some qualitative researchers may remain epistemologically reticent to reuse, arguably a more pressing issue relates to the structural challenges facing qualitative researchers seeking to deposit and use data from archives. As Weller (2023) notes, the development of archival infrastructure has been accompanied by the establishment of guidance, procedures, and practices for data sharing and reuse. While there are benefits to researchers and research participants in effective data management practices and the protection of data, increased bureaucracy and regulation may be experienced as constraining or stifling. Examples include regulatory frameworks such as the General Data Protection Regulation (GDPR), which Mauthner (2012) suggests has focused on procedural matters at the expense of considering the emotional and intellectual investments necessary for knowledge creation, curation, and reuse. This issue is of significance in the neo-liberal academy, where there remains a tendency to prioritise quantitative methods and data management practices (Weller, 2023). At the same time, academic casualization serves to negatively affect the capacity of researchers—often moving between short term and part-time contracts—to accommodate expectations for preparing data effectively for preservation.

Archiving and the process of gaining consent to archive might also impact on the data itself. This is more problematic given the requirement from certain funders to place data from research that they fund in the archive, thereby making researchers obliged to ask for consent to archive. Participants, for instance, may be unwilling to take part because of the request to archive, some interviews may take place ‘off the record’, while others will require pseudonymisation (replacing personal data with artificial identifiers) and anonymisation (removing personal data so it is no longer identifiable) (Heaton, 2004, p. 81; Parry & Mauthner, 2004, p. 146). While redaction and anonymisation is ethically and legally important, the process of ensuring anonymity might “require removing so much detail that the data are rendered meaningless” (Parry & Mauthner, 2004, p. 144).

We are clear that engagement in reuse requires careful consideration, especially in thinking about the role of context and how the absence of background information about the data might affect understandings (we discuss this further in Chap. 4), and how it relates to more recent concerns around the capacity of staff in the neo-liberal academy. However, the new methods and new sources of data available for reuse undoubtedly present significant opportunities for qualitative research and the knowledge it can produce. Mason (2007) argued that investing in the qualitative archive would afford more opportunities for thinking and examining social change from a distinctive qualitative perspective. In particular, she saw the qualitative archive project as creating the possibility of “scaling up” through data sharing, to produce cross-contextual understandings and explanations (Mason, 2002, p. 4). While we concur that archives and reuse can better enable us to fully exploit the potential of qualitative research and the data generated, the approach we take is not a one-way process which moves from small scale to big, or from intensive to extensive levels. Rather, the approach moves iteratively, integrating both extensive and intensive analytical levels.

3.6 How to Search an Archive

Now that you are equipped with knowledge about the key available archives, and where to find them, your task is to begin exploring the data available in one or several selected archives identified as of interest. Your aim is to undertake an initial search to acquire a precursory understanding of the nature, quality, and focus of the available data sets and their 'fit' with your secondary analytic research topic. As part of this initial identification of data sets it can also be useful to explore the publications and outputs produced by the original researchers. This process may be wide-ranging, locating data sets on a general topic area. So, to take an example, you might begin your search with a broad keyword, such as 'adoption' or 'workers' rights'.

Alternatively, you may already have narrower parameters for your search, with a focus on searching for data to fit a specific substantive issue and set of research questions. Searches could instead consider specific aspects of adoption, such as 'kinship care', or a specific area of workers' rights, such as the 'digital economy'. Additional parameters, such as geography or date of data collection, might be added to your search to further narrow your results. A point to remember is that if you begin with tight or prescriptive search parameters, you may need to be prepared to expand your search should it return no results, or extend your keywords to include alternative terminologies, expressions, or synonyms. The approach taken, and decisions therein, should remain tightly linked to your overall research.

Many archives will allow Boolean operators, which enable you to narrow and expand your search terms. Table 3.2 shows examples, although bear in mind these may vary, with different sites providing more or different operators. Some archives will also provide support in conducting comprehensive keyword search. The UK Data Service collection makes use of *The Humanities and Social Science Electronic Thesaurus* (HASSET). This is a working tool to help you retrieve the data from the

Table 3.2 Boolean operators

Boolean Term	Description
AND	AND will narrow your search results to include only relevant results that contain your required keywords. Example: "crime" AND "poverty"
' '	To search for phrases, type the phrase in quotation marks. For example, type "kinship care" in quotes in the search box
OR	Use 'OR' when you only need at least one of the search words to appear in the results. Example: "poverty OR disadvantage"
NOT or -	Use 'NOT' or '–' when you do not want a particular term to appear in the results. Examples: "heart defects NOT attack" or "heart defects -attack"
*	Use '*' as a wildcard to search on all words that contain the letters you enter. You must enter a minimum of three letters plus the '*' character. Example: "fam*" would find "family" or "families"
+	Use '+' when you require the exact word to appear in all the results. For multiple words, you must put a + sign in front of each word that must be exact. Example: search on "+park dogs" if you only want to retrieve results with dogs in parks

UK Data Service collection which best relates to your field of research or interest (https://vocabularies.ukdataservice.ac.uk/hasset/en/).

A further point is that although in some cases you may be able to download data immediately, many online archives require researchers to register and sign an agreement or licence before they can access and download data sets. Usually these 'end user' terms and conditions include undertaking not to use the data for commercial purposes; to keep the data secure and only share it with registered users; to preserve confidentiality and not attempt to identify individuals, households, or organisations in data sets; and to destroy all copies of the data once registered use is completed. We discuss the ethical issues associated with data preservation and reuse in Chap. 8.

3.6.1 Accessing the Archive—Issues to Consider

It will become apparent as you begin your search that each archive is structured differently, operating through a unique interface, and with its own specific procedures for searching its database. Many archives will include information on 'how to use' the service. Timescapes, for example, includes an interactive demonstration. The UK Data Archive, similarly, has a wealth of documents on how to access the data and its terms and conditions. Whichever archive(s) you choose, first examine the available documentation to help guide your search. Archives typically include a function which allows users to search the catalogue. The UK Data Service (UKDS) search function allows users to filter specifically data sets with qualitative sources, as well as topic, country, and date. The Finnish Social Science Data Archive has grouped its data sets by type (qualitative and quantitative) and theme. Further filters, such as date and keyword, can also be used.

An issue for most archival search functions is that the keywords are those provided by the original researchers when they deposited their data set. This means that there may be data that is of interest in archived material that is not returned by a descriptive word search. So, looking again at the example of adoption as a broad search term, in a study on inner city family life there may be useful cases within a data set which discusses adoption. However, if this was not a key topic identified by the original researchers, the data set may not include this as a keyword. In some archives it also may not be possible to filter for qualitative data, thereby protracting the search strategy. For specialist or community archives, like the GALA project in South Africa, you may find that no searchable archive or keywords exist, but rather data is stored under individual projects. Accessing data in this way will require research on the individual projects and their published outputs to determine what data is actually stored, its relevance, and how (and if) it might be accessed. It may also be possible to contact members of the original research team to determine whether the project might contain material suitable for your investigation (we discuss this further in Chap. 8).

Advances in digitisation are making positive in-roads into these issues by digitising all data content and connecting these data to search functions. This means that a

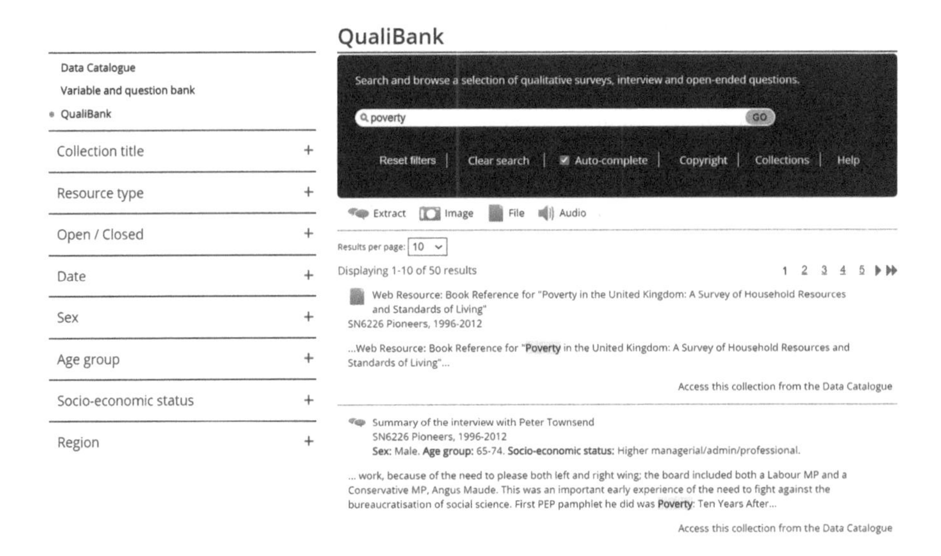

Fig. 3.1 Example of content search in Qualibank. (Source: https://discover.ukdataservice.ac.uk//QualiBank)

simple search will not only scan keywords, but also document content. Timescapes, the archive we used for our project, has a function that allows users to search within specific projects, as well as individual cases. Such searches can then be filtered by criteria such as gender, employment status, age group, year of birth, and ethnicity. Qualibank is also continuing to digitise all its content, with the ability to search all content currently only available for certain collections. In Fig. 3.1, we see an example from the Pioneers Study, where the search term ‘poverty’ is identifying specific interview transcripts (in this case an interview with sociologist Peter Townsend) where the word ‘poverty’ is mentioned.

A final issue to consider is whether there actually is any relevant secondary data available for your purposes. Our own project was developed with a specific interest in gender and care across time. We were already aware of the Timescapes Data Archive, with some members of the research team having been involved in the original research. Although archives have developed, it may not be possible to locate data sets on the topic you wish. There are a number of avenues you can take at this point. As well as re-examining your keywords, you may also consider extending your search parameters to include data sets in different locations, or for different dates. Alternatively, through your knowledge of the field of investigation, you may know that there are researchers who have collected or are collecting data on your topic of interest who have not (yet) archived their data, and it may be possible for them to share their material with you. Such a strategy may be ethically problematic as consent may not allow researchers to share data beyond the original study. However, it may be possible, in some cases, to revisit participants and update consent to allow data sharing. This process is likely to be time-consuming, speculative, and therefore unfunded. Where such an endeavour might be more practical is where research teams working with data on similar substantive topics form formal data sharing agreements. Such an approach depends significantly on researchers having existing networks within the field, and the presence of relationships that would support such an agreement being put in place.

If these strategies are unproductive, you may take this as evidence that there is a gap in this area of research, thereby justifying primary data collection. By working either outside or alongside the archive, new, creative possibilities for extending the scope of big qual methods can also be considered.

3.6.2 Data Quality and the Politics of the Archive

The 'quality' of qualitative data was discussed in Chap. 2 and is an issue that follows us into the archive. As discussed, archives set out rules that govern what data is included, and excluded, from the archive. Indeed, it is quite possible for a data set prepared by a researcher to be rejected by an archive if it fails to meet established thresholds for quality controls. A further key issue is that repositories do not always have the capacity and resources to ingest data sets. So, it is not unheard of for a researcher to be under an obligation to prepare a data set for deposit, only to find that the archive is unable to process the collection at that point, if at all.

For data to be considered by an archive, there are typically protocols and procedures in place which establish criteria to determine the 'quality' of the collections and place controls on access. Quality in this context will likely relate to the extent to which the data is well curated, well documented, and adequately referenced and indexed (Corti et al., 2020, p. 13). These are also the key criteria upon which the next stage of the breadth-and-depth method relies, since if the data collections are poorly recorded, or if there are inconsistencies or inaccuracies, it makes it time-consuming, if not impossible, to assemble the data into a useable format. One concern is that guidance on quality is frequently created with quantitative data in mind, and therefore neglects the complexities involved in qualitative archiving.

While systematic ordering, description, and recording is crucial in being able to find and use data suitable to our needs, quality will not necessarily be heterogeneous within or across archives, and as such any endeavour should take account of the history of the archive and its archiving procedures. It is therefore important to consider and assess the quality of the data, and to reflect on how it might affect your own project. This is not to say that the intended research cannot proceed, but it should commence with an awareness of any possible caveats and limitations in the resultant analysis. We will return to the issue of quality in more depth in Chap. 4, but during any initial search of an archive and preliminary consideration of possible data sets, you may use the following questions to guide you:

- Who created the data and for whom? Is there any identifiable conflict of interest?
- What is the purpose of the data being collected? Does that purpose match with your interests? If not, is that important?
- How comprehensive are the accompanying documents, such as participant information sheets, consent forms, and interview schedules?

- Does the data seem to be of reasonable quality? Is it in an accessible format?
- Does the archived data appear complete, and are there any identifiable gaps? For example, is there a summary of the data collected? If so, does this match with material in the archive? Is there any information about data that has been withheld, or redacted? Are there are metadata files which can tell you about the characteristics of the data collected?

Rasmussen and Blank (2007) have argued that data reuse and re-analysis is enhanced by increasing the 'quality' of the data available. They align quality to processes of systematisation, which force the original researcher to rigorously document and record the data being stored. In the context of reuse of research data, then, ideas or perhaps even ideals of quality are connected with a certain view of what an archive is and its function. While archives have moved beyond what Popple et al. (2020) describe as the traditional and monolithic view of archiving and record keeping, certain assumptions remain in terms of who we expect to use the archive and why. Vardigan and Whiteman (2007) have looked at the question of who uses and gains access to the archive, and supports the need to expand this to a much wider group of users.

Quality for archived data, of course, is not just systematically labelling and recording data but more profoundly reflects sets of practices of collection or co-production, of discard or preservation. Even with systematic processes in place, we should also remain aware that archived data retains a 'trace' of the bureaucracy of ethics committees and associated procedures that informed what data was collected in the first place and how. Moreover, the archive itself and its own processes can be shaped by cultural understandings and political perspectives on what type of data is considered important or of value, or by the funding and support for data collection efforts. As noted, in the example from Burundi in Case Study 3.1, the collection of both qualitative and quantitative data has been shaped by political processes of data collection aligning to political ideologies.

3.7 Outside and Alongside the Archive

We wish to end this chapter with a final note about data that might be considered as sitting outside or alongside community or formal archives. The earlier part of this chapter provided a long list of types of qualitative data, and in your own search you may find yourself heading towards that which is familiar for qualitative researchers: textual material solicited through interviews. This is especially the case for those researchers working in the field of qualitative research.

However, big data, as Chap. 1 was keen to show, has produced a wealth of new materials and technologies. These can enable us to access a much larger variety and volume of data that we might have otherwise not considered in a qualitatively driven project. One of the most significant forms of data emerging as potential research material is data scraped from the web. This might include Twitter and other social media data, and in Chap. 1 we referenced a number of studies working with this type of material. Digital materials are constantly expanding, with opportunities available not just through social media, but other digital spaces such as blogs, forums, and discussion boards. These spaces generate the possibility of creating your personal archives of scraped data which can be analysed alone or combined with other qualitative data sources.

Digitisation has also opened up the possibility of using the breadth-and-depth approach to analyse open responses to large-scale surveys. Digital surveys that are building in open-ended responses may be single point in time, collected as a one-off for a specific project, or longitudinal panel surveys, such as the UK's Understanding Society or US's Fragile Families and Child Wellbeing initiatives, where open-ended questions are inserted into waves of data collection at certain points. All these forms of material can be expanded further with the possibilities of combining multiple data sets into a new assemblage of data.

It is the possibility of bringing data sets together that is the greatest advantage of the breadth-and-depth method, and we have seen several alternative relational configurations of data emerging in recent years. In Neufeld et al. (2022) the breadth-and-depth method was employed by pooling 11 data sets from ten studies conducted in different food environments in Bangladesh, Cote D'Ivoire, England, Ethiopia, India, Indonesia, the Gambia, and South Africa (see also Chap. 6 and Case Study 6.2). Purcell et al. (2020) similarly used the breadth-and-depth method to pool data from multiple studies sharing the same broad topic, whilst retaining the attention to detail and depth of analysis that characterise good qualitative research. For this project, five data sets, comprising one-to-one interviews with 138 women who had undergone abortion, were used. Rather than access data from the archive, these were all studies previously completed by the research team (Case Study 4.4). Other examples include the case studies involving projects combining multiple data sets on HIV prevention (Case Study 3.5) and large-scale international projects which link survey and qualitative data over time (Case Study 4.1). Other approaches which can inform big qual endeavours include Hughes et al.'s (2022, p. 375) "continuous, collective, and configurative approaches to qualitative secondary analysis". Their work skilfully brought together research teams from two separate studies. Although the study remained 'small scale' in terms of the number of cases considered, its methodological and ethical approach provides valuable insight into the benefits of incorporating a collective and participatory ethos into data reuse.

3.8 Moving Your Search Strategy Forward

This chapter has reviewed archives and archiving, and shown the different ways in which the qualitative archive has developed internationally. As well as providing possible starting points for your project, it has emphasised the need to exercise caution as you enter the archive, being aware of the cultural, social, and political settings in which the data was preserved. We close with a number of areas that are worthy of contemplation at this stage which may help guide your thinking:

- In what ways might your search strategy shape your data set selection?

- What caveats and limitations should you bear in mind working across different archives?
- What new possibilities may emerge from working alongside or outside the archive?

3.9 Resources

You can find a full list and map of archives internationally at the UK Data Archive: https://ukdataservice.ac.uk/get-data/other-providers/data-archives.aspx

The 'Data in Brief' journal comprises papers that outline accessible data sets: https://www.sciencedirect.com/journal/data-in-brief

Case Study 3.1 Burundi's Statistical Visa: Coordinating or Controlling Data Collection? *Jean-Benoît Falisse and Godefroid Bigirimana*

Data extractivism is the practice of researchers coming to a context, collecting data using local labour, and disappearing. It is rife in low-income and so-called 'fragile' settings (Marchais et al., 2020), which are also regularly plagued by the duplication of (costly) large-scale data collection efforts by non-governmental organisations and researchers (Altay & Labonte, 2014). Poorly coordinated and unaccountable qualitative and quantitative data collection initiatives are a significant hurdle for the development of research on often crucial aspects of social, political, and economic life in some of the world's most challenging contexts. Burundi's 'statistical visa' came, in part, in response to these issues.

The visa is effectively an authorization to carry out research that needs to be obtained from the Technical Committee for Statistical Information [CTIS]. The policy was introduced in 2013 as part of a reform seeking to increase the transparency and availability of data on the country and its population. The scale of the data collection, rather than the nature of the data collected, is in theory the key criterion for deciding whether a research project requires the visa. The application requires submitting research instruments and plans—a practice that is not dissimilar from the national ethics clearance processes of many countries (incidentally, such ethical clearance is also a requirement of the statistical visa). On paper, the reform had the potential to create large qualitative and quantitative data sets: the National Statistical Institute [ISTEEBU], which works closely with the CTIS (the head of ISTEEBU is also the head of CTIS), started listing visas on its website and was, therefore, ideally placed to signpost and connect research data sets. By late 2021, however, only a few projects were listed, and even fewer data sets were archived. As often, the resources accompanying the policy were quite limited. More importantly, however, the visa was hijacked by politicians.

By mid-2015, Burundi was not the model of liberal peace-building that had been heralded for a decade anymore, it was turning into an oppressive and totalitarian state where information is controlled (Falisse & Nkengurutse, 2019). Suspicions that the visa was always more about control than regulation and archiving were voiced from the onset of the project, but the late 2010s made the situation even more explicit. The visa could be used to veto politically-sensitive projects and became a powerful deterrent from doing research: only six visas were granted in 2015–2017. What is more, ISTEEBU, which has a 'fast track' visa application (the normal route takes months when not over year), has used such privilege to establish a quasi-monopoly on large-scale research in the country. As of 2021, more visas were being issued but their nature had also changed on a crucial detail: whereas earlier visas only requested the applicant to transmit the final version of their report, new visas also request the full data set. The effort to create large, public, data set has moved from a coordination effort from ISTEEBU where researcher would share the raw data if they felt comfortable, to a tightly controlled and mandatory process. Whether this will allow the most necessary large-scale collection of qualitative and quantitative data on the sensitive topics of livelihoods, violence, and governance is, at the very best, uncertain.

References

Altay, N., & Labonte, M. (2014). Challenges in humanitarian information management and exchange: evidence from Haiti. *Disasters*, *38*(s1), S50–S72.

Falisse, J. B., & Nkengurutse, H. (2019). From FM radio stations to Internet 2.0 overnight: information, participation and social media in post-failed coup Burundi. In *Social media and politics in Africa: Democracy, censorship and security*. Zed Books.

Marchais, G., Bazuzi, P., & Amani Lameke, A. (2020). The data is gold, and we are the gold-diggers: Whiteness, race and contemporary academic research in eastern DRC. *Critical African Studies*, *12*(3), 372–394.

▶ Case Study 3.2 Qualitative Research Data in Germany with Qualiservice *Jan-Ocko Heuer*

In Germany, archiving and sharing materials from qualitative research is not as common as in the UK, as researchers are usually not obliged by research funding organizations to do so. However, it is becoming increasingly more popular. One main service provider for qualitative data sharing is the Research Data Center [RDC] Qualiservice at the University of Bremen. While some specialized qualitative RDCs in Germany exist (for an overview: https://www.konsortswd.de/en/datacentres/all-datacentres/), Qualiservice is the only RDC that archives and provides qualitative research materials from across the social sciences—independent of the (sub-)discipline, research topic, or data type. Qualiservice offers a broad range of services to primary researchers and data users. This includes study-specific advice on data preparation throughout a research project and many information materials, templates and tools (also in English) on informed consent, anonymization, and documentation of research data (see here: https://www.qualiservice.org/en/the-helpdesk.html).

Qualiservice is the successor of the "Archive for Life Course Research" [*Archiv für Lebenslaufforschung*, ALLF], which emerged in the early 2000s from the Collaborative Research Center [CRC] 186 "Status Passages and Risks in the Life Course" at the University of Bremen (project term: 1988–2001). The CRC had examined the relationships between social structures, social changes, life-course patterns and individual biographies using both quantitative and qualitative data. The qualitative data of the CRC were taken over by ALLF, creating an original corpus of about 700 interview transcripts that were digitised, anonymized and prepared for secondary uses. In the mid-2000s, ALLF conducted a feasibility study (in cooperation with the main RDC for *quantitative* data in Germany, the GESIS Leibniz Institute for Social Sciences, and funded by the German Research Foundation DFG) to examine the potential demand and supply for a data service center for qualitative interview data and to develop conceptual foundations for an RDC. From 2011 to 2014, Qualiservice received funding from the DFG to develop workflows for archiving and sharing qualitative interview data in close collaboration with the scientific community. In 2015, Betina Hollstein, Professor of Qualitative Methods and Microsociology at the University of Bremen, took over the management of Qualiservice and integrated it into the SOCIUM Research Center on Inequality and Social Policy.

In 2019, Qualiservice was officially accredited as an RDC by the German Data Forum [RatSWD], which monitors the quality of all accredited data centers. Data sharing is organized together with the certified World Data Center PANGAEA and the State and University Library Bremen [SuUB]. For distributed archiving of mixed-methods-studies, a workflow has been developed with the GESIS Leibniz Institute for Social Sciences. Moreover, in cooperation with the Scientific Information Service Social and Cultural Anthropology [FID SKA], Qualiservice has created workflows for archiving and sharing data from ethnographic research (e.g. field notes, observation protocols, research diaries, photos, or audio data). Today, Qualiservice hosts research materials from a broad range of studies, encompassing written materials, visual and audio-visual data in several languages. Qualiservice is committed to the FAIR Guiding Principles for scientific data management and stewardship.

The future of data sharing in Germany will be shaped by the National Research Data Infrastructure [NFDI], a recent large national initiative with the aim of creating a coordinated network of discipline-specific consortia that offer various data services to research communities to improve data sharing both nationally and internationally. As one of the members of the "Consortium for the Social, Behavioural, Educational and Economic Sciences" [KonsortSWD], Qualiservice is coordinating the development of "QualidataNet", a federated infrastructure for sharing qualitative research data in Germany—this prepares the way for fostering a culture of data sharing in qualitative research.

For further information on Qualiservice, visit the website https://www.qualiservice.org/en/ or contact Qualiservice via info@qualiservice.org

▶ Case Study 3.3 The Timescapes Data Archive Archive *Kahryn Hughes, Bren Neale, Graham Blythe, Brenda Phillips, and Rachel Proudfoot*

The Timescapes Data Archive is a specialist resource of Qualitative Longitudinal Research [QLR] data, developed under the ESRC Timescapes initiative, the first large scale qualitative longitudinal study to be funded in the UK by the ESRC (2007–2012). Part of the remit of Timescapes was to build a resource of QLR data for sharing and re-use, in a context where relatively few such resources were available at that time. The Archive is built on a model of disaggregated preservation and comprises rich digital, multi-media holdings of QLR data, and forms part of an institutional repository at the University of Leeds, UK (Research Data Leeds). The Archive was developed so that it operates as a satellite repository for QLR data to the national archive, the UK Data Archive. It conforms to international archiving standards and data are copied to the UKDA for longer term preservation. Data deposit and registered use are managed through robust legal, ethical, and contractual processes.

Holding over 3000 files of qualitative data, the Timescapes Data Archive, provides a unique opportunity for digital delivery for the secondary analysis of qualitative data. QLR data sets deposited in the archive are 'by formal agreement on a restricted access basis'. Anyone can discover the Archive's data resources via its website but only approved re-users from an academic/policy institution can access the data. This involves completing an application to access data sets of interest, providing reason(s) for seeking access, and signing a contractual assurance to uphold the data set and archive's re-use Terms and Conditions. Once an application is reviewed and approved by the

Timescapes Director, a user login is created providing time-limited access to the Archived data sets. The Timescapes Data Archive website hosts an extensive 'knowledge bank' of cutting-edge methods resources. These comprise a growing corpus of work on methods of QLR research and qualitative data reuse, as well as data management guidelines for QLR study design, and research protocols such as consent for archiving agreements.

Since its development, the Archive's functionality has expanded. The 'Big Qual: Breadth-and-Depth Method' data set in the Timescapes Data Archive, has been specifically designed as a specialist teaching resource by National Centre for Research Methods [NCRM] researchers and is now also available for tutors to access. This is an internationally-accessible, free-to-use, resource, that supports learners to "gain hands-on experience of accessing, searching, obtaining, and organising qualitative longitudinal data; build learner practice in identifying and exploring data sets using contextual metadata; and grow learner skills for making in-depth analyses of context, complexity, and details as present within the data.' Interested researchers and tutors skilled in the use of archives have been able to successfully deliver teaching objectives by developing modules of study focussed on the themed data sets hosted by the Archive, as well as outputs from such teaching. The Research Data Management Service actively underpins this tutor-led activity through the creation and delivery of its 'Timescapes for Teaching' resource, developed to support tutors in a classroom and/or online setting. This resource allows tutors to assign individual student and group access to the Archived Timescapes data for educational/and or research purposes, whilst reinforcing the security, governance, and ethical aspects of data reuse.

The Timescapes Data Archive offers endless and exciting possibilities for analysing data through time, over the life course and across the generations, whether conducting depth analysis of small data samples or scaling up to conduct big qual analysis across amalgamated data sets.

Case Study 3.4 Working Across Data Sets in the Irish Qualitative Data Archive: An Example and Future Possibilities *Jane Gray*

Hosted by Maynooth University Social Sciences Institute, the Irish Qualitative Data Archive [IQDA] was founded in 2008 under the Irish Government's Programme for Research in Third Level Institutions. The archive was established as a central access point for all qualitative social science data generated in and about Ireland. It also had the task of framing standards and protocols for sharing and re-using qualitative data in Ireland. Subsequently, IQDA was a founding member of the Digital Repository of Ireland [DRI], which was officially launched in

2015. The DRI is a national digital repository for Ireland's humanities, social sciences, and cultural heritage data. Data deposited in IQDA are preserved and disseminated by the DRI. In this case study, we discuss the *Family Rhythms* project as an early example of working across data sets deposited in IQDA. We then describe the potential for future big qual research with qualitative data sets in the DRI.

Family Rhythms was funded in 2012 by the Irish Research Council as a demonstrator project for re-using archived qualitative social science data. With the aim of developing a comprehensive understanding of family change in twentieth century Ireland, the project brought together 100 biographical interviews from the Life Histories and Social Change [LHSC] project with qualitative interviews with parents and children from 116 families that participated in the first wave of the National Longitudinal Study on Children (Growing Up in Ireland [GUI]). Working across the data sets presented both opportunities and challenges. Most interestingly, they opened different temporal windows on family change. LHSC adopts a retrospective perspective on the lives of three birth cohorts of people born at different times in twentieth century Ireland. By contrast, the GUI interviews have a prospective perspective as part of an ongoing panel study. We addressed this challenge through a strategy that we have described as 'working backwards and forwards' across the data, comparing changing family experiences across different life course, generational and historical standpoints. For example, to understand changing grandparent—grandchild relationships we began with the GUI child interviews as a contemporary reference point, worked 'backwards' in time comparing memories of relationships with grandparents across the LHSC cohorts and then 'forwards' comparing grandparents' relationships with grandchildren.

The DRI provides significant opportunities for social science researchers to use innovative big qual approaches in constructing and analysing data assemblages, taking advantage of similarities and differences between data sets to address new questions. Researchers might seek out similar data to expand scope and range. For example, there are many oral and life history studies in DRI—not all of them within IQDA—that provide an opportunity to include diverse experiences, such as those of people who experienced life in institutions. Alternatively, researchers might explore a topic such as children's voices by creating an assemblage incorporating a wide range of data forms (essays, drawings, interviews) collected for different research purposes. To assist researchers in identifying data, the DRI uses standardised metadata and recommends the use of standardised vocabularies and the inclusion of contextual data.

References

Geraghty, R. & Gray, J. (2017). Family rhythms: Re-Visioning family change in Ireland using qualitative archived data from growing up in Ireland and life histories and social change. *Irish Journal of Sociology, 25*(2), 207–213.
Digital Repository of Ireland: https://www.dri.ie/
Irish Qualitative Data Archive: https://www.maynoothuniversity.ie/iqda

Case Study 3.5 Saving and Reanimating a Classic Data Set: The Women, Risk and AIDS Project *Rachel Thomson*

The Women, Risk and AIDS project was a landmark feminist social research project—a part of a programme of research funded by the UK Economic and Social Research Council to better understand the sexual cultures and practice of the UK population in the face of the threat of HIV/AIDs which was at that time an untreatable and terminal diagnosis. The WRAP project was involved 150 interviews with young women aged 16–21 living in Manchester and in London—and was one of the first research projects to use in-depth tape-recorded interviews as a method for exploring young women's sexual identities, values, and practices within a critical feminist framework. The research was highly generative at the time and became influential in shaping a generation of interview-based feminist social research. Conducted before norms were established around data archiving or preservation, the transcribed interviews were in the possession of the original research team. New research funding in 2018 enabled a research team to digitise, anonymise and archive the data set, collecting relevant contextual documentation (including oral history interviews with the original researchers) and producing metadata enabling the collection to be found and accessed through digital research techniques. Alongside this process researchers also engaged in a project to share the materials with contemporary audiences in ways that might capture interest and create new meaning. Examples of this work include the development of a drama project with a student theatre of society based on 30-year-old interviews with drama students at the same University; creative work with young women at youth clubs and memory work with former youth-workers and parents. The interviews have also been shared with academic audiences, utilised by historians interested in social change and social scientists interested in interviews and qualitative data analysis. In line with the team's interest in building intergenerational feminist communities a collaborative blog and exhibitions have been created thorough the open access platform Omeka that provides access to the collection. In developing and sharing the archive the research team have engaged in careful ethical work, honouring a promise to preserve anonymity while also realising the value and integrity of the interviews, and investing in the places and

communities that they form part. Key ideas explored by the team have included what it means to 'reanimate' archives as well as ideas of 'rematriation' that involve reuniting data with the communities that generated them and engaging in 'careful ethics' that may include working without explicit consents for data sharing. To find out more about the project visit:

Our project blog and website http://reanimatingdata.co.uk/

Our Omeka site that showcases the FAYS collection (Feminist Approaches to Youth Sexualities) https://archives.reanimatingdata.co.uk

The archive https://sussex.figshare.com/collections/Women_s_Risk_and_Aid_Project_Manchester_1989-1990/4433834/2

The Reanimating Data project team (2018–2021) includes: Rachel Thomson, Niamh Moore, Sharon Webb, Ester McGeeney and Rosie Gahnstrom and was funded by the ESRC Ref: ES/R009538/1

The Women, Risk and AIDS Project team (1988–1990) includes: Janet Holland, Caroline Ramazanoglu, Sue Sharpe and Rachel Thomson

References

Bastain, J. A., & Flynn, A. (Eds.). (2018). *Community archives, community spaces: Heritage, memory and identity*. Facet Publishing.

Besek, J. M., Coats, J., Fitzgerald, B., Mossink, W., LeFurgy, W., Muir, A., Rasenberger, M., & Weston, C. (2008). International study on the impact of copyright law on digital preservation. *International Journal of Digital Curation, 3*(2), 103–111.

Bishop, L., & Kuula-Luumi, A. (2017). Revisiting qualitative data reuse: A decade on. *SAGE Open, 7*, 1.

Chauvette, A., Schick-Makaroff, K., & Molzahn, A. E. (2019). Open data in qualitative research. *International Journal of Qualitative Methods, 18*.

Cook, T. (2013). Evidence, memory, identity, and community: Four shifting archival paradigms. *Archival Science, 13*, 95–120.

Corti, L. (2000). Progress and problems of preserving and providing access to qualitative data for social research: The international picture of an emerging culture. *Forum: Qualitative Social Research, 1*(3).

Corti, L., & Thompson, P. (2004). Secondary analysis of archived data. In C. Seale, G. Gobo, J. F. Gubrium, & D. Silverman (Eds.), *Qualitative research practice* (pp. 297–313). Sage.

Corti, L., Van den Eynden, V., Bishop, L., & Woollard, M. (2020). *Managing and sharing research data: A guide to good practice*. Sage.

Doorn, P., & Tjalsma, H. (2007). Introduction: Archiving research data. *Archival Science, 7*(1), 1–20.

Duchesne, S., & Brugidou, M. (2016). BeQuali—An archive in question: Looking back at the creation of a qualitative data archive. *Revue d'anthropologie des connaissances, 4*(4), o-an.

Elman, C., Kapiszewski, D., & Vinuela, L. (2010). Qualitative data archiving: Rewards and challenges. *PS: Political Science and Politics, 43*(1), 23–27.
Elsesser, K. M., & Lever, J. (2011). Does gender bias against female leaders persist? Quantitative and qualitative data from a large-scale survey. *Human Relations, 64*, 1555–1578.
Evans, J. (2018). Indigenous archiving and wellbeing: Surviving, thriving, reconciling. In J. A. Bastain & A. Flynn (Eds.), *Community archives, community spaces: Heritage, memory and identity*. Facet Publishing.
Fink, A. S. (2000). The role of the researcher in the qualitative research process: A potential barrier to archiving qualitative data. *Forum Qualitative Sozialforschung / Forum: Qualitative Social Research, 1*(3).
Heaton, J. (2004). *Reworking qualitative data*. Sage.
Heaton, J. (2008). Secondary analysis of qualitative data: An overview. *Historical Social Research., 33*(3), 33–45.
Hughes, K., & Tarrant, A. (Eds.). (2019). *Qualitative secondary analysis*. Sage Publications.
Hughes, K., Hughes, J., & Tarrant, A. (2022). Working at a remove: Continuous, collective, and configurative approaches to qualitative secondary analysis. *Quality & Quantity, 56*, 375–394.
James, J. B., & Sørensen, A. (2000). Archiving longitudinal data for future research: Why qualitative data add to a study's usefulness. *Forum Qualitative Sozialforschung/Forum: Qualitative Social Research [Online journal], 1*(3).
Kluge, S., & Opitz, D. (2000). Computer-aided archiving of qualitative data with the database system "QBiQ". *Forum Qualitative Sozialforschung / Forum: Qualitative Social Research, 1*(3), Art. 11.
Kuula, A. (2000). Making qualitative data fit the "data documentation initiative"; or vice versa? *Forum Qualitative Sozialforschung / Forum: Qualitative Social Research, 1*(3).
Mason, J. (2002). *Qualitative research resources: A discussion paper*. Prepared for the ESRC Research Resources Board. Unpublished, obtained from author.
Mason, J. (2007). "Re-using" qualitative data: On the merits of an investigative epistemology. *Sociological Research Online, 12*(3), 1–4.
Mauthner, N. (2012). Are research data a common resource? *Feminists@Law, 2*(2), 1–22.
McLeod, J., O'Connor, K., & Davis, N. (2020). *Doing research differently: Archiving & sharing qualitative data in studies of childhood, education and youth*. The University of Melbourne McLeod-OConnor-Davis_Doing-Research-Differently_Discussion-Paper.pdf (hasscloud.net.au).
Moore, N., Salter, A., Stanley, L., & Tamboukou, M. (2017). *The archive project: Archival research in the social sciences*. Routledge.
Mozersky, J., Parsons, M., Walsh, H., Baldwin, K., McIntosh, T., & DuBois, J. M. (2020a). Research participant views regarding qualitative data sharing. *Ethics and Human Research, 42*(2), 13–27.
Mozersky, J., Walsh, H., Parsons, M., McIntosh, T., Baldwin, K., & DuBois, J. M. (2020b). Are we ready to share qualitative research data? Knowledge and preparedness among qualitative researchers, IRB members, and data repository curators. *IASSIST Quarterly, 43*(4), 952.
Neufeld, L. M., Andrade, E. B., Ballonoff Suleiman, A., Barker, M., Beal, T., Blum, L. S., Demmler, K. M., Dogra, S., Hardy-Johnson, P., Lahiri, A., Larson, N., Roberto, C. A., Rodríguez-Ramírez, S., Sethi, V., Shamah-Levy, T., Strömmer, S., Tumilowicz, A., Weller, S., & Zou, Z. (2022). Food choice in transition: Adolescent autonomy, agency, and the food environment. *The Lancet, 399*, 185–197.
Parry, O., & Mauthner, N. S. (2004). Whose data are they anyway? Practical, legal and ethical issues in archiving qualitative research data. *Sociology, 38*(1), 139–152.
Popple, S., Prescott, A., & Mutibwa, D. H. (2020). Community archives and the creation of living knowledges. In S. Popple, D. H. Mutibwa, & A. Prescott (Eds.), *Communities, archives and new collaborative practices*. Policy Press.
Purcell, C., Maxwell, K., Bloomer, F., Rowlands, S., & Hoggart, L. (2020). Toward normalising abortion: Findings from a qualitative secondary analysis study. *Culture, Health & Sexuality, 22*(12), 1349–1364.

Rasmussen, K. B., & Blank, G. (2007). The data documentation initiative: A preservation standard for research. *Archival Science, 7*, 55–71. https://doi.org/10.1007/s10502-006-9036-0

Simionica, K. (2018). Self-documentation of Thai communities: Reflective thoughts on the Western concept of community archives. In J. Bastian & A. Flinn (Eds.), *Community archives, community spaces: Heritage, memory and identity* (pp. 79–96). Facet.

Smioski, A. (2011). Archiving qualitative data: Infrastructure, acquisition, documentation, distribution. Experiences from WISDOM, the Austrian data archive. *Forum Qualitative Sozialforschung / Forum: Qualitative Social Research, 12*(3), Art. 18.

Thompson, P. (1991). *Pilot study of archiving qualitative data*: Report to ESRC, Department of Sociology, University of Essex.

Vardigan, M., & Whiteman, C. (2007). ICPSR meets OAIS: Applying the OAIS reference model in the social science archive context. *Archival Science, 7*(1), 73–87.

Weller, S. (2023). Fostering habits of care: Reframing qualitative data sharing policies and practices. *Qualitative Research, 23*(4), 1022–1041.

Wright, L. H. V., Tisdall, K., & Moore, N. (2021). Taking emotions seriously: Fun and pride in participatory research. *Emotion, Space and Society, 41*.

Ziebland, S., & Hunt, K. (2014). Using secondary analysis of qualitative data of patient experiences of health care to inform health services research and policy. *Journal of Health Services Research & Policy, 19*(3), 177–182.

'Aerial Surveying': Overviewing the Data and Constructing a Corpus

4

4.1 Introduction

The previous chapter gave an overview of the potential qualitative data sources that can be utilised in the breadth-and-depth method, and how to source and search for them. As well as data preserved in archived collections, we highlighted the diverse range of alternative materials that might be gathered in this process. Having now located data sources, and identified several possible data sets, we can move onto the 'aerial survey'. In this step we describe ways to gain an overview of your selected data, with the aim of constructing a corpus that consists of multiple data sets. The steps that follow are not prescriptive, and we would encourage you to employ flexibility depending on the nature and form of your selected projects or data sets. The case studies provided throughout illustrate different approaches to combining data and handling metadata.

To summarise, in this chapter we will cover:

- a brief overview of the metaphor 'aerial survey'
- data auditing, including an overview of 'metadata'
- issues relating to good data management and storage
- approaches to building your new corpus
- the challenges and limitations you may face

4.2 The Aerial Survey

As introduced in Chap. 2, the breadth-and-depth method uses the metaphor of archaeology to think about large qualitative data sets as a landscape. A useful visualisation of step one—the aerial survey—is the view one has from a reconnaissance aircraft (see Fig. 4.1). As you look down you may see various marks, textures, and features on the ground below. At this initial stage, we do not go to the surface and begin an investigation of individual sites. Rather, our objective is to gain an

© The Author(s), under exclusive license to Springer Nature Switzerland AG 2023

S. Weller et al., *Big Qual*, https://doi.org/10.1007/978-3-031-36324-5_4

Fig. 4.1 Archaeological metaphor and step one of the breadth-and-depth method. (Illustration created by Chris Shipton https://www.chrisshipton.co.uk/)

overview of the characteristics of the landscape and identify potential features of interest for further exploration.

The starting point is to identify possible data sources that you might consider in your own breadth-and-depth project. If you have not already selected your data sets for analysis, Chap. 3 provides a number of examples and places to begin your search. A preliminary exploration might involve a detailed search of multiple archives to determine data sets of interest. Such an exploration could include reflection on the different types of data sets that you have identified and their relationship to and utility for your research topic.

Once you have a list of possible data sets you wish to take forward, you can begin your aerial survey. This entails a systematic audit of your material to determine whether it is of an appropriate nature, quality, and 'fit' with your research topic. You should also consider at this point whether it would be possible to reconstruct the selected data into one large corpus. This process would apply regardless of your sources of data. The data sets under consideration might be multiple data sets curated in one archive or data sets from different archives. They might also be a combination of archived data sets and your own existing data, or other material. Or indeed, you may choose to select from within data sets and work only with a subsample of the materials available, for example using data generated via a particular method or sampling using participant characteristics. Tarrant's project, detailed in

Case Study 4.1, theoretically sampled from two of the Timescapes qualitative longitudinal data sets. The studies—Intergenerational Exchange (IGE) and Following Young Fathers (FYF)—were brought into conversation to build a picture about how, why, and for whom these men were engaging in care.

In our project, we worked with six data sets from the Timescapes Data Archive. These sets of data were from *different* but *linked* empirical projects. Members of our team had been involved in two of the original Timescapes projects, and were therefore personally connected to the material, and aware of the nature and form of the files archived. Nonetheless, as a team we were not familiar with the content of the other projects, and, more importantly, we were yet to understand how—if at all—we might seek to bring the data from all six projects together for the purposes of our new study. From the outset, researchers employing the breadth-and-depth method will have different relationships to the data they select. In some cases, such as our own, you may have an existing connection to the data and therefore inhabit an 'insider position'. For instance, you may know or have been part of the original research team, and be party to knowledge about project decision-making and/or have direct experience of data generation and analysis. Others may inhabit more of an 'outsider' position knowing little or nothing about the data. Attempts to connect to the original research team may not be possible. This positioning might also shift over time as you become more familiar with the data. Neither position should be considered 'better' or more valuable. While an outsider position can bring fresh or alternative insights, an insider position can provide a more intimate understanding of the context of the studies under examination. Regardless of what a researcher's position is, the aerial survey is the starting point from which new relationships with the data can be formed.

Whether you have knowledge of the data you plan to use, before you begin your analysis it is necessary first to determine whether the selected data sets can be meaningfully reconstructed into a new data assemblage. Both quality and heterogeneity across the data sets are significant when making this assessment. Data sets may initially look appropriate for your purposes, but it is not always immediately possible to determine their format or quality. Access to some of the selected data may be restricted as it is deemed sensitive, for instance. Some data may be embargoed for a set period and not available for reuse in the immediate future. Other data may be poorly labelled or missing without reason. These variations can impact on the extent to which data sets can be combined meaningfully. Thinking both within and across your data sets, Table 4.1 provides a jargon-free list of questions to ask as you begin your data audit.

4.3 Understanding Metadata and Meta-Narratives

To answer these questions and determine if and how your selected data sets could be combined, we draw on metadata and metanarratives. Metadata, put simply, is documentation (or data) that describes data, or as it is often termed 'data about data'. A basic example of metadata is the information stored alongside an electronic

Table 4.1 Questions for researchers auditing their data

• What type of data has been stored, and how well is it labelled? • How much contextual information has been curated alongside the project and the individual materials? • Does it include what you need to know to judge whether the data may be a promising fit with your aim(s) (e.g. summary of the substantive focus and aim(s) of the research, the methods used, a description of locality/setting(s), details of the timeframe, an outline of the characteristics of participants, the unit(s) of analysis employed, and details regarding the sampling frame)? • Is any of the data embargoed, restricted, or missing? If so, might this impact on your analysis? • Are the selected data sets from different disciplines, or philosophical or methodological traditions? How might any differences impact on your ability to address your research questions? • How much data management, reorganisation, and active epistemological thinking is required to address these issues, and enable the data sets to be combined?

document such as the type of file, the date the file was created and/or last modified, and its size. This information can help you organise your files and work with your data more effectively. The same is true for research data. Metadata can help us navigate a project and is especially useful for preserving its long-term useability and meaningfulness during reuse (Faniel & Yakel, 2011). Researchers may record, and make available, significant metadata about the study. Within qualitative data, characteristics or attributes attached to individual files or cases might include social and economic descriptors such as age, income, class, and ethnicity. This contrasts to quantitative data where such categories are the data, while the metadata would be the survey instruments used to measure these variables.

Metanarratives are a form of metadata that can include more descriptive information about the make-up of the original research team, for example documents which detail the research aims, design, methodology, and disciplinary background. These provide an understanding of how, when, and why a study was conducted, and help to situate the study by providing context that might otherwise be unknown to the secondary researcher. Attributes attached to data sets can extend to information about the study's geographical location, temporal coverage, data collection method, recruitment and sampling, the observation unit (the case), and ethical procedures.

For archivists, metadata and metanarratives not only aid the identification, location, and interpretation of records, but they also enable the archive to be used more effectively by researchers in perpetuity. As Kuula (2000, p. 3) notes, they can act as a "bridge between the original collector and the re-user giving the essential information for secondary analyses". By providing these details, the original researcher is not only helping others locate suitable data but is also supporting reuse and interpretation. For researchers employing the breadth-and-depth method, scrutiny of significant contextual narratives and associated lists of metadata provides a precursory understanding of the nature, quality, and suitability of data sets. From there, we can identify analytical units that have a 'fit' with our own research questions (Hammersley, 2010) and use them to assemble a new corpus. As we will describe later in the chapter, metadata has an important role in allowing you to pursue your logic of inquiry (see Chap. 2).

Dodd's research, which is detailed in Case Study 4.2 (see also Dodds et al., 2021), used an aerial survey to assist in identifying and selecting possible data sets for analysis. Here, eight social scientists formed a collaborative network, with the aim of bringing together large volumes of qualitative data on HIV and biomedicalisation in the UK. Focused on a pivotal moment in HIV research, an initial list of 20 studies was reviewed for potential analysis. This involved each team considering the metadata and metanarratives for their own projects and reporting back to the network. This aerial survey allowed studies to be discounted due to, for example, irrelevance, data availability, and ethical considerations. In this case, a total of 12 data sets involving 589 individual participants were identified for inclusion in the new data assemblage.

If you are accessing all or part of your data from formal archives, it is likely that at least some descriptive, structural, or administrative metadata will be available. While the metadata recorded with each research project will vary (see Chap. 2 and the resource section of this chapter for examples), there are several materials that you can expect to find alongside the data. These are listed in Table 4.2. Of course, not all studies will deposit the full range of materials even when they seem pertinent to the research. In some instances, metadata may be excluded from the archive due to participants being identifiable (e.g. details about the study location) or consent having not been obtained (e.g. field notes, participant diaries, and visual data). As we will discuss, gaps may also simply reflect inaccurate or inconsistent data collection practices during the study.

One of the advantages of accessing data from formal archives is that the metadata preserved has typically passed through a set threshold of quality control. However, as you conduct your aerial survey, it is worth bearing in mind that this will vary across archives, collections, and data sets (see Chap. 1 for a discussion). Corti (2007, p. 2) has also highlighted the lack of "agreed practical procedures for preparing, storing and disseminating qualitative data" across qualitative archives. At the

Table 4.2 Materials and metadata commonly archived

Materials and metadata
Data files—this could include audio or video recordings, transcribed files, photographs, drawings
Individual data—files may themselves have metadata attached such as interview ID, age, sex, gender, class, occupation, location, place of interview, date of interview
Study aims and objectives
Study geographical and temporal coverage
Code book, with any variable names, codes, definitions, and abbreviations
Headnotes (a brief summary of the material that follows)
Field notes or reflective diaries written by the researcher
Project research materials: topic guide/interview schedule
Ethical procedures: consent forms; details about data processing and preparation for storage.
Information on research methods, for example description of the method used, recruitment, and sampling
Publications or reports from the research

time, she observed that within qualitative data archives there were "few common descriptive standards, access to many collections is poor, and there are no integrated resource discovery tools" (2007, p. 37). Weller (2022), writing more recently, points to the continued dominance of quantitative data management strategies within archives, and the difficulties researchers face in adapting these standardised approaches to qualitative research.

Organisations such as the UK Data Archive (UKDA) and the Consortium of European Social Sciences Data Archives (CESSDA) have been instructive in developing qualitative data management procedures, which, in turn, has enabled qualitative archives to gain credibility and commitment from funding bodies. Nonetheless, not all projects—especially older data sets—will match the standards established by, for example, the UKDA. The context of limited funding and investment can prevent archives from implementing robust data management and storage procedures. Improving standards, notably, does not necessarily mean digitisation but rather improving knowledge about and physical access to archives. As you commence an overview of your selected data sets, it is important to bear these challenges in mind.

It is worth noting that alternative considerations come to the fore where data is being accessed from alternative sources. For example, individual researchers working on similar substantive topics may wish to bring their data sets together. Such a process is not straightforward, and original agreements relating to consent and data sharing must be considered before data sets are pooled. An exemplar of this type of work was executed by Dodds et al. (2021) and is detailed in Case Study 4.2. Here, the network of researchers worked closely to develop their own anonymisation protocol. Only once this protocol was applied to each data set was any data shared out with the original research teams.

4.3.1 Using Metadata to Audit Your Data Sets

Having gained a broad understanding of what materials are available for your data sets of interest, you can begin to audit the content. At this point, the data sets are not merged but remain in their original form. The information you record will depend on your approach to sampling, analysis, and the study design. Crucially, the amount of delving required to uncover this information will depend on the extent and nature of information that a particular data archive requires.

In our project, this process involved auditing the materials archived for the six data sets of interest. Using short summary tables for each project (example is below in Fig. 4.2) we recorded information on:

- the number and type of files available (i.e. pseudonymised transcriptions, visual materials, field notes)
- details of any embargoed materials (i.e. audio files, sensitive material)

Project 1: Sibling and Friends

- **Additional materials available** (e.g. schedules, base data, project guide): Deposited but none apparent
- **Total number of data files:** 606 currently in archive (+ missing files known to have been deposited + embargoed audio).
- **Details of any embargoed data:** All audio files and part of Jay's Wave 3 transcript
- **Number of individuals/families:** 44 cases – up to 52 young people as this includes group interviews
- **No. of Waves:** 3 (2002-5, 2007, 2009)
- **Geographical spread:** UK-wide
- **Keywords assigned to each case (not individual files):** Male/Female (as appropriate); Case; Childhood; Family; Siblings; Brothers; Sisters; Friendship; Help; Schooling; Education; Employment; Hobbies; Memories; Rules; Responsibilities; Place; Feelings; Generational Identities; Inter-Generational Relationships; Historical Events; Local; Global; Past; Present; Future
- **Base data:** Comprehensive spreadsheet; quite a lot of detail around parents' jobs/family circumstances – use with care so as to not reveal identities.

Case ID	WAVE 1 (2002-05, HERITAGE)				WAVE 2 (2007)						WAVE 3 (2009)			
	Transcript	Circle map	Survey	Activity sheets	Transcript	Notes	Circle map	Timeline	Diary	Interim tasks[1]	Transcript	Notes	Circle map	Photo(s)
Alisha[2]	✓	✓			✓	✓	✓	✓			✓	✓	✓	
Allie	✓	✓			✓	✓	✓	✓			✓	✓	✓	✓1
AshleyB[3]	✓		✓		✓	✓	?	?			✓	✓	?	✓1
Bethany	✓			✓5	✓	✓	✓	✓			n/a	n/a	n/a	n/a
Chelsea and Emma	✓				✓	✓	✓2	✓2			✓	✓	✓2	
Cora	✓				✓	✓	✓	✓			n/a	n/a	n/a	n/a
Danielle	✓			✓2	✓	✓	✓	✓			✓	✓	✓	✓3
JazzyB	✓		✓6		✓	✓	✓	✓			✓	✓	✓	✓2
Jim	✓	✓		✓1	✓	✓	✓	✓			n/a	n/a	n/a	n/a
Kelly, Jessie and	✓	✓2		✓8	✓	✓	✓3	✓3			✓	?	✓3	✓4

? = deposited but missing from list in archive
[1] Postal activities – 'cultural commentaries' and 'my life @ 25'
[2] Incorrectly labelled 2008 in Alisha – should be 2003
[3] Dates not accurate on archive labelling

Fig. 4.2 Example from Timescapes Data Archive audit

- the number and type of cases (i.e. the units of analysis constructed by the original research teams such as individual or family group)—and because our data set was longitudinal this included participation in each wave of data collection
- the metadata (i.e. descriptive, structural, and administrative) and availability of contextual materials (i.e. field notes, project guides, or reports)
- sample characteristics including date of birth and gender
- the geographical spread of the sample
- the keywords that the original researcher assigned to their data set as a whole to enable archival searches

Auditing the data in this way provided us with a strong sense of the scope and nature of the data including format and volume, research tools used, the substantive emphasis (also given by published papers and data collection instruments such as interview schedules), and an understanding of the temporal rhythm of data collection including the spacing of each wave of data. In short, we documented the 'who', 'what', 'where', 'when', and 'how' of the original research endeavours. This activity also, crucially, revealed anomalies in the landscape such as gaps in contextual material or unusual, misplaced, or incorrectly labelled or categorised files. For example, we identified a number that had not been archived in one of the projects. In most cases, missing or edited data was clearly identified as having been embargoed for ethical reasons, such as consent or confidentiality. In others, it was not possible in the position of 'outsider' to understand why an item was not in the archive. For example, was an interview not there because that person missed a wave, or because it had not been archived, or had the content been embargoed for some reason?

These issues are not unusual, nor are they unique or distinctive to the Timescapes Data Archive, but rather they reflect archiving practices more generally. Crow and Powell (2010) point out that what qualitative data is stored and recorded, and in what form, reflects research practices, and the wider social, cultural, and institutional setting in which the study was conducted. Temporality also shapes the form and content of the data, as data management and storage practices have changed significantly over time, especially with the introduction of digitisation. What is considered 'missing' might therefore be different across space and time. Prompted by Crow and Powell (2010), it is also useful to question for whom data is collected and for whom it is stored (and therefore for whom it is missing). For example, there may be data not formally collected because a digital recorder fails, or because a participant continues to talk after the recorder stops. That knowledge may be absent from the archive, but it remains with the participant and the researcher. It is also the case that what is included in the archive is a reflection of the original research team, and what they consider the most important information to archive. A missing part of the conversation might be included in a field note, or it may not. The data and metadata archived is also determined by the temporal resources to which researchers have access. As we discuss in Chap. 8, archiving qualitative data is a time-consuming, labour-intensive activity, often inadequately funded and resourced. This can impact on the data management practices implemented.

As noted, in our project we already had a partial 'insider' positioning. This allowed us, in some cases, to be able to contact the original research team to check the information collated or query any anomalies. This dialogue allowed us to gain greater insights into the projects, and arguably gain a deeper insider perspective. We were also able to check our understandings and interpretations and determine the reasons for any gaps and inconsistencies. As detailed in Case Study 4.1, Tarrant's (2017, p. 604) research also engaged with the original researchers via collaborative meetings and extended reflective dialogue. As well as providing insight into the context of the source study, Tarrant drew on knowledge of the original researchers to help identify subsamples within the data that might be especially useful to her work. While the types of conversations described in our work and Tarrant's can be helpful, their absence does not preclude gaining understanding and insight. First, you may decide that ontologically you prefer to retain an outsider position and focus instead on building your own relationship with the data. Second, it is an issue of pragmatism. Such conversations may simply not be possible—the original researchers may be uncontactable, for example because they have moved employment or retired. Furthermore, if the selected projects are relatively old, researchers may have forgotten the details you require.

In our data audit we reviewed all the material that had been preserved, with the aim of gaining strong contextual understanding of the selected projects and the data stored. Our substantive intellectual interest related to gendered care and intimacy across the life course, and as such we undertook the audit with the specific objective of identifying metadata that would allow us to assemble a new data corpus that would enable us to examine this social phenomenon. Crucially, then, the data audit cannot be regarded as an administrative or organisational task. As Mason (2018,

p. 193) points out, thinking about how you want to organise, search within, and manipulate your data is an epistemological question, deeply connected to your relationship to the phenomena you wish to study. Shortly, we will discuss how we decided to reorganise our data in a manner most fitting to addressing our research questions. First, and drawing on our own experience, we consider both the practical and intellectual challenges associated with working with metadata.

4.4 Working with Metadata—Challenges

4.4.1 Closeness and Distance

We begin by addressing an issue well rehearsed in the secondary qualitative data field, that of context and the distance that researchers reusing data might have from the original project. While much discussed, this issue has remained pertinent to the breadth-and-depth method and merits brief reflection here. Discussing the development of the CESSDA, Smioski (2011) noted that a key challenge in establishing qualitative archives was the reluctance (at the time) of some qualitative researchers to support them. As noted elsewhere (see Corti, 2000; and also Heaton, 2019 for an overview), this caution centred on the perceived relationality between qualitative researchers and primary data. The argument presented is that context is an experience only primary researchers can have, and that such 'closeness' is central to understanding. From this standpoint, qualitative secondary analysis (QSA) not only lacks contextual insight but is at risk of misinterpretation. During early presentations of our planned approach to merge multiple sets of archived qualitative data we occasionally encountered similar responses. 'Why would you want to do that?' was a question put to us by some qualitative researchers. Working with large volumes of data could only amplify the absence of context and relationality.

While some researchers remain reticent, there is undoubtedly a growing interest in, and appreciation of, qualitative data and its reuse. This has been energised by the emergence of big qual as a new field of research. Supported by robust critiques from realist and constructivist positions (Hammersley, 1997, 2010; Moore, 2007), scholars have been keen to highlight that the absence of contextual data cannot be considered as unique to QSA. Doing so creates a false distinction between primary and secondary data by failing to acknowledge that all data are subject to a process of recontextualisation and reconstruction, constitution, and re-constitution.

Mason (2007) helpfully notes that certain forms of interpretation and explanation are, in fact, only possible with distance. As a research team, we had different relationships, and levels 'closeness', to the data with which we were working. Significant contextual and descriptive metadata, as described by Kuulla (2000, p. 3), provided us with 'a bridge' to the original research. This bridge enriched our understandings of the projects being considered. However, there were also areas where we remained outsiders—we had not 'been there' to collect the data in most of the projects. And while there were practical and intellectual challenges associated with this, distance offered a range of potential gains, most notably the ability to observe continuity and change in substantive topics across different data sets.

4.4.2 Data Harmonisation

Despite assumptions that secondary analysis is more time and resource efficient than generating new data, in our experience, working with multiple data sets was laborious, particularly in the early stages of the project. While metadata can provide a 'bridge' into a study for reuse, the attributes that the original researchers preserve will primarily reflect *their* interests and research questions. Your data audit might reveal, therefore, epistemological differences across your selected data frames or in how the substantive focus of the research is conceptualised.

These differences may shape the project metadata preserved. You may, for example, be seeking to combine multiple studies related to youth crime. The studies under consideration may approach and define this issue differently, depending on their disciplinary and epistemological stance. Research into youth crime, for example, will be approached differently by sociologists, psychologists, or urban planners. Likewise, the researchers' epistemological understanding of 'youth' may be distinct, and therefore shape how the study is conducted. A second, but related, distinction connects to definition and categorisation. To continue our example of youth crime, you may define 'youth' as young people up to the age of 21. However, your chosen data set(s) may not record 'youth' in as detailed or nuanced a way as you might wish. For example, they may not have recorded participants' date of birth, or they may have used age cohorts that do not correspond with your own work. Or researchers may assign terms like 'adolescent', as in developmental psychology, which conflict with other more sociological and socially constructed ideas of 'youth'.

One such example in our project related to metadata on family type. Within this step, we were keen to explore family type as a possible means of informing theoretical sampling and aiding comparative analysis. As we ventured into the archive, we found that there was no coherent means of categorising 'family' across the six projects. Rather, information was recorded about who the participant(s) lived with. This information was gathered qualitatively, often through a combination of questions asked at interviews and material derived from interview data. Despite a level of coherence, we found that the data recorded often did not identify key relationships that we felt might be important for our theoretical sampling. For example, we were lacking baseline data that took account of different family types, such as foster children, stepfamilies, grandchildren who lived with grandparents, and friends. Categories such as 'other' rarely provided detailed information. Overall, we concluded that the baseline data was not sufficiently detailed when it came to family structure for our purposes of considering how family types may influence care and intimacy. Our own experiences also point to the wider challenges of defining and categorising aspects of participants' identities with contextual and metadata.

A final point to consider is that you may also be working with data sets that encompass both quantitative variables and qualitative data. As detailed by Crivello in Case Study 4.3, The Young Lives study includes a wealth of data, ranging from Panel Survey data to school surveys and in-depth qualitative longitudinal interviews. This large data set presents opportunities for linking numerical survey data with qualitative data. Yet as Crivello notes, the ability for researchers to effectively and meaningfully ‘link’ different types of data depends on the quality of data management and the ethics protocols in place. In this example, data shared a coding framework which enabled research questions to be explored within and across data sets. Not all projects will have anticipated the possibility of linking data, making linkages difficult if not impossible.

With these issues in mind, you want to ensure that through your audit you make yourself aware of differences within and across your selected data. You may decide that the data is recorded in a manner inappropriate for your intended analysis, and therefore not possible to use. Alternatively, you may devise mechanisms and practices that will allow you to manage them in your analysis. This may include combining or merging categories into something workable or selecting a subset of a project sample so as only to include certain types or forms of data (e.g. only interviews). This process of auditing, reflection, and subsequent data management can be time-consuming, especially should you decide not to use the data set. Whichever approach you take, missing or inconsistent data should not necessarily prevent reuse. It may well be that digging deeper into the material will provide the information you require. A critical reflection on the possible impact on study design can be considered alongside alternative approaches to determine the way forward. Incorporating an element of pragmatism in your decision-making is also important since the data you use, whether from an archive or another source, will never ask questions in precisely the way you might wish them to be asked.

4.4.3 Constructing your Corpus

The careful combination of multiple qualitative data sets can meaningfully strengthen claims about understanding how social processes work. Since big qual is a relatively new methodological endeavour, there are limited exemplars of how to go about the process of corpus construction. Recent work has centred on combining thematically linked data sets. As noted, Tarrant’s substantive research aim was to investigate men’s care responsibilities within low-income families (Case Study 4.1). The project devised was to include QSA as a distinct stage of research. This involved bringing two studies on similar substantive issues from the Timescapes Data Archive—‘Following Young Fathers’ and ‘Intergenerational Exchange’—into analytical conversation with each other and new primary research. The dialogue between the data from the two studies was able to build upon and extend the interpretations of the original research teams. In Wilson’s (2014) research, secondary data was combined with primary data as a tool for reflexive sociology. Wilson had already produced a significant body of qualitative data which had involved children and young people characterised as 'vulnerable'. She specifically sought archived qualitative data exploring the same issues as her own (e.g. sibling relations, friendships, family life) but with a different (in her words 'more ordinary') sample. Combining data from two distinct but thematically linked data sets enabled Wilson to reflect on the ways in which concepts such as 'vulnerability' had been framed in different research contexts. In Edwards and Cabellero's (2010) study, data from lone mothers of mixed race children was extracted from a larger lone mothers sample in Marsden's classic study. This was then innovatively brought into dialogue with accounts from contemporary mothers of mixed race children. In Purcell and Maxwell's study of abortion stigma (Case 4.4) a new data assemblage was created by merging 11 unique qualitative data sets. Interestingly, they found that the new data set supported the comparison of women's accounts of abortion. It was the comparison of attitudinal data, which was deeply connected to context, that required more careful consideration.

To assemble a new corpus, consideration must also be given to your proposed analytic unit (or units). Different data sets may speak to similar substantive questions or aspects of social life but may have collected data from different sources, and in different forms. The typical analytic unit in qualitative research is at the level of an individual person. Analytic units can also relate to family units, households, a neighbourhood, organisations, programmes, and so on. Several different analytic units can also be 'nested' within each other, and used within and alongside each other in a single study.

Within our project, we initially assumed that we would take the individual person as our basic unit of analysis. However, during our audit we found a variety of forms of data. Most materials were based on repeat interviews with the same individual. Others included both 'primary' participants and 'significant others', often grandparents or siblings. Further variations included group interviews with siblings and multiple members of a family. One of the consequences of this variety was that the quality and type of metadata attached to individuals varied across data sets. While these differences were not sufficient for rejecting the projects identified, they prompted reflection on whether an individual research participant should be treated as our analytic case, or whether it was necessary to examine alternatives. This issue of what comprises a case is revisited in Chap. 7.

4.5 Assembling Your New Corpus

4.5.1 Using Metadata to Create a New Corpus

The final part of this step is to construct your new corpus. How you do this will depend on the metadata identified during your audit, and how it relates to your substantive topic of interest. Pragmatism is key here since the data and the metadata will likely never provide a perfect fit with your logic of inquiry. You may, for example, have to make compromises in the age cohorts you use, or work with categorisations of social class that you wouldn't use to conduct primary research. Time is a further complexity since it is likely that the data sets being brought together will not have been collected at the same point in time. Similarly, if you are using projects that collect data longitudinally, synchronic or cross-cutting analysis will be difficult since the timing of the waves of interviews across projects will not necessarily correspond.

In our project, our aim was to move away from a comparative analysis by project to create a new assemblage which could address our research questions related to gendered practices of care and intimacy over time. The Timescapes Data Archive presented a unique source since the curated projects covered key stages of the life course: childhood, youth, 'middle' age, and the oldest generation. Given our interest in the life course, we were initially keen to organise the data with reference to time in some dimension. After reflecting on the metadata collected within each project, and the different dimensions of time captured, we decided to reorganise by gender and age cohort, both analytic units that would allow us to examine gendered practices of care and intimacy and their change over time.

Having already audited the research material, the organisational process was relatively straightforward. The metadata on year of birth and gender of interviewees was used to label the pooled data into age cohorts of men and women. Once complete, the constituent data became one entity: the corpus for our project. Whilst the assemblage was organised by age and gender cohorts, each data file retained links to the original metadata. A key question was how to construct our age cohorts. As shown in Table 4.3, these were primarily theoretically formed around critical economic, social, and demographic trends identified as having had a significant impact on care, intimacy, and personal relations. However, in order to create analytically

Table 4.3 Age cohorts by birth year, social change, and sample by gender

Born	Examples of social change related to intimacy and care	Female	Male	Total
1908–1949	Experience of First and Second World Wars, rationing, and austerity Pre-comprehensive welfare state Defined male and female roles	13	5	18
1950–1969	Comprehensive welfare state with security from 'cradle to grave' established following 1948 National Insurance, National Assistance, and National Health Service Acts. Low unemployment and social housing provision Rise in female emancipation, start of availability of contraceptive pill, change in attitudes to sex and marriage, 1967 Abortion Act	24	23	47
1970–1989	Rising employment rate and protection for women: 1970 Equal Pay Act, 1975 Sex Discrimination Act, 1975 Employment Protection Act Rising divorce rate and cohabitation: 1971 Divorce Reform Act	14	16	30
1990–2001	Extended dependency in childhood with longer time in education Later age of marriage and having children Experience of austerity with public service cutbacks, and health and care privatisation	28	27	55

workable categories, we had to make small adjustments to our selected age cohorts to ensure there was a sufficient number of cases in each age band.

While our categories were generally operational, we also had to make pragmatic decisions that worked with our data. Such decisions relative to sampling or the recategorisation of metadata decisions should be acknowledged clearly in your research design. As you work through different options, you may find that differences that initially seem limiting afford new perspectives and analytical insights. Data collected across different time points, for example, can allow you to consider change and continuities in the substantive topic being investigated.

4.5.2 Data Management

Data management and organisation is critical as you begin to combine your data sets. This may mean, for example, creating new labels for all your data to ensure you have workable, harmonised file names. At a simple level, this aids retrieval, and enables the quick reorganisation of files. However, such considerations have epistemological significance since they can determine the shape and nature of your data sets, and the analytical processes which follow. In our research, we devised a system where the file names provided a clear sense of the source and nature of the data. For example, from the file names we were still able to determine the original source project, and which interview wave and type of data it related to. It is also worth noting that clear file names, and attached metadata, are also a requirement of the computational software used in the next stage of the method. This is discussed in Chap. 5.

Storing your data in a safe, secure space is important for all research, and it is especially important when working with multiple sets of data from different sources. If you have accessed data from an archive, it is important to ensure that you have read any data sharing and privacy agreements that relate to data use, and that you are aware of any requirements set by the archive. These could relate to how you store the data, how long you retain it, and how you dispose of it. Systems of data transfer and storage that enable encryption are recommended. It is also worth bearing in mind that large volumes of qualitative data can require significant storage space, especially if audio or visual materials are included. Cloud-based options can be problematic in terms of data security and may breach the terms of archive agreements, as well as institutional ethical approval.

As with all aspects of this method, it is vital that you develop a system that works for your big qual project. This can be especially important if you are working collaboratively across teams in different institutions. Before embarking on your project, we would recommend you discuss the options available with relevant departments in your own organisation.

4.6 Conclusion

This chapter has provided an outline of the aerial survey. At this stage, our aim is to gain an overview of the characteristics of the landscape of our data, and identify potential features of interest for further exploration. This involves a systematic audit of your material to determine whether it is of an appropriate nature, quality, and 'fit' with your research topic. We have discussed the importance of using metadata to audit your data sets, but also highlighted the possible limitations and challenges. This might include missing data or other gaps in knowledge, or having to reconfigure data to allow data harmonisation. We have also reflected on our own relationship to data as secondary analysts, and the significance of closeness and distance. Finally, we examined ways in which you might assemble your new corpus using metadata. Bear in mind that this approach is not linear, and the connections between steps are iterative. As such you may still, at the end of this stage, move back a step to examine alternative or additional data sources.

4.7 Resources

It is useful to look at examples of guidance for transcription, storing study description, and metadata sheets. Examples are available from the Timescapes studies: https://timescapes-archive.leeds.ac.uk/wp-content/uploads/sites/47/2018/04/Transcription-guidelines-and-model-23July08-current.pdf

The Finnish Social Science Data Archive holds a wide range of resources on data description and metadata: Data Description and Metadata—Finnish Social Science Data Archive (FSD) (tuni.fi)

Further examples are available from research from the public space art project called "My Public Living Room" in Toronto. Here you will find a data set guide and data example: FSD2999 My Public Living Room Interviews 2014 | Aila (tuni.fi)

Case Study 4.1 Building New Empirical Research Using Archived Qualitative Data; 'Men, Poverty and Lifetimes of Care'

Anna Tarrant

The 'Men, Poverty and Lifetimes of Care' (hereafter MPLC, 2014–2018) study funded by the Leverhulme Trust (ECF-2014-228) employed a novel two-stage research strategy to explore men's care responsibilities and support needs in low-income contexts.

The first stage included engagement with, and secondary analysis of, theoretically sampled data from two qualitative longitudinal data sets stored in the Timescapes Data Archive to produce new research questions and refine theories for investigation in a new phase of empirical work. The two studies were Intergenerational Exchange

[IGE] and Following Young Fathers [FYF]. IGE investigated the support and care provided by vulnerable and marginalised grandparents with an analytic focus predominantly on the experiences of grandmothers, despite the inclusion of grandfathers. FYF explored the parenting journeys and support needs of a diverse sample of young fathers (aged 25 and under), including a sub sample of ten participants who were identified as marginalised. Given the orientation of FYF, there was more limited analysis of the interdependencies of these young fathers with other generations of men. Working within and across both data sets, the QSA work for MPLC involved initial analyses of how men narrated their care responsibilities. As this work progressed, it was possible to build a picture about how, why and for whom these men were engaging in care. Apprehension of the processes and contexts that either constrained or enabled their participation in family contexts was also possible (Tarrant, 2017).

This work also informed a second phase of new data generation with a sample of young fathers, mid-life fathers and men engaged in intergenerational and informal and formal kinship care arrangements for children who were also family members (Tarrant, 2021). While affording some substantive insights, there were specific issues raised by both generations of men that were not explored, followed up or analysed in depth by the original research teams (e.g., around child maintenance payments). In addition, further analyses confirmed that it was not possible to fully understand familial intergenerational relationships *between* the two intragenerational cohorts that comprised the new sample (Tarrant and Hughes, 2020). This made it difficult to support rigorous substantive claims through the QSA work. Nevertheless, it was possible to build new, participant driven questions into the next empirical phase of the study and to construct a sample more reflective of the diversity of men's participation in low-income families across the lifecourse.

MPLC demonstrates that the re-use of existing qualitative data is a fruitful and viable method for identifying and tackling research problems, especially those that are otherwise little examined or explored in existing academic literature.

References

Tarrant, A. (2017). 'Getting out of the swamp': A realist secondary analysis of several qualitative longitudinal data sets to develop research design, *International Journal of Social Research Methodology*, 20(6), 599–611.

Tarrant, A., & Hughes, K. (2020). Qualitative secondary analysis: Building longitudinal samples to understand men's generational identities in low income contexts. *Sociology, 53*(3), 538–553.

Tarrant, A. (2021) *Fathering and poverty: Uncovering men's family participation in low-income contexts*. Policy Press.

Case Study 4.2 How a Team of Researchers Undertook an Initial Exploration of HIV and Biomedicalisation Across 12 UK Qualitative Data Sets

Catherine Dodds

In 2015, a network of eight social scientists came together to identify what would be needed to collaboratively re-analyse and archive a substantial amount of qualitative data spanning several decades. We came from a range of different institutions and were highly familiar with one another's' work on the social, structural and behavioural issues relating to HIV in the UK. The discussion about qualitative data re-use was stimulated by a Wellcome Trust Seed grant during a key moment in the HIV epidemic, when the technical race for the 'AIDS' finish line was picking up pace with the success and repurposing of HIV treatments increasingly promoted as the key to ending AIDS in a generation (UNAIDS, 2014). This rapid bio-technological shift in the use of pharmaceuticals represented an entirely new framing of the AIDS experience; contrasting sharply with experiences among researchers and people impacted by HIV after 1996 when effective treatments first emerged. Among social scientists, the concepts of biomedicalisation (Clarke et al., 2003) and pharmaceuticalisation (Bell & Figert, 2015) had started to offer fruitful critical insights to our joint reflections on this period of time, and we held these concepts in mind while deciding what data sets might help to afford an 'aerial view' of the terrain. By bringing together data from projects situated directly within this timeframe, we anticipated that this pilot work would help to bring new insights about the future prospects for HIV Treatment as Prevention as one that was likely to continue to be impacted by social and geographic stratification, uncertainty and inequality, just as initial and ongoing treatment roll-out had been. (For more details see the full-length paper on our project in Dodds et al., 2021). A further aim of this work was to see if we would be able to make these and other routes of enquiry more feasible for future researchers, by exploring the possibility of archiving multiple qualitative data sets that had been traditionally stored by individual researchers and their institutions.

Initially, members of the network discussed and shared a list of our own relevant UK projects undertaken between 1997–2013—the period during which HIV treatments had been introduced, refined, expanded and re-purposed—to consider which data sets would hold information on the implications of HIV antiretrovirals for health, risk and policy-making. We settled on a long list of just under twenty potential studies and asked all members of the network to collate and assess the availability of study materials such as transcripts as well as metadata, research protocols, demographic descriptions of each sample, topic guides, interview schedules and fieldnotes. This process facilitated the development of a core list of potential data sets with a brief description of each. We sought to include all qualitative projects that network members had led in that timeframe that directly related to living with diagnosed HIV; and/or reflections on risk among those not diagnosed, and/or projects relating to an array of HIV services or policy-making. We all agreed that the mobilisation of this opportunistic sample of researchers and our work would be the most feasible means of getting this type of qualitative data re-use off the ground in our field, as it was unlikely to happen any other way. During the collation of study materials, individual transcripts and sometimes entire projects were discounted from inclusion either due to irrelevance, technical issues (file corruption) or where sharing potentially identifiable life history narrative material was deemed inappropriate. Our network ultimately settled on 12 data sets that involved contributions from 589 individual participants via a range of iterative exchanges.

Throughout this work, it was essential for everyone in the team to acknowledge the very human task of assembling data as it travels on 'data journeys' across time, space and groups of researchers (Leonelli, 2016; Hammersley, 2010). During our first steering group meetings (consisting of all network members contributing data sets, and additional key advisers including a historian of public health and a policy adviser from a national HIV third sector organisation) we discussed the emotional labour of recollection as we re-opened old files, and throughout our further interactions with these materials we remained considerably alert to both the challenging and also the mundane aspects of lives lived in close proximity to HIV (among research participants and researchers alike). In addition, as the project continued, we noted the importance of listening out for silences and missing voices in the data sets and project lists, which itself can enable new analytical avenues (Irwin et al., 2012).

A key learning moment that supported us in readying the materials for our first analytical 'sweep' was our collaborative development of anonymisation protocol that was required in order to help us with the initial sharing of transcripts within our own network. This protocol ensured we removed and replaced direct identifiers (such as personal names/addresses) using a standardised format. It also helped us to agree on a code of practice in dealing with indirect identifiers such as geographic location/place of employment—with a shared commitment to retain as many of these as possible, but only doing so while ensuring that a string of indirect identifiers could be potentially used to identify an individual. It was the development and operationalising of this protocol which helped us to recognise what a sizeable and skilled task the anonymisation of all 12 full data sets would be, before these data could be shared further within the network, or deposited for archiving with the UKDA.

Ultimately, the contributing researchers randomly sampled 3 transcripts/notes from each of the data sets they had contributed in order to test and further refine the anonymisation protocol. That enabled two members of the network to gain access to and work with 36 datapoints from across the 12 studies to undertake an initial framework analysis, thematically coding all materials that related to HIV antiretroviral medication (and its absence) as well as the effects this had on lived experience or decision making.

Despite the small volume of coded material in relation to the overall data held within the combined data sets, we were astonished by the richness of enquiry that this entry-level analysis of 36 diverse transcripts afforded. It helped us to understand in much greater depth which particular studies we might want to bring into more direct conversation with one another in future re-analysis, as well as future routes of enquiry for the broader re-use of all 12 data sets. During our final Steering Group meeting, we brought these emergent themes into focussed discussion by using a 'data workshop' approach (Tarrant, 2016)—which helped us to explore not only the contents of the data, but the surprises that they held, and the contexts of data production in rich detail. This project was initiated as a means of testing the feasibility of data re-use, and in turn it helped us to see how much we had underestimated the labour that would be needed to ultimately archive all these data sets, while also clarifying the numerous lines of enquiry that were afforded by bringing them together for data re-use.

Case Study 4.3 Young Lives: Linking Large Volumes of Survey and Qualitative Data across Time

Gina Crivello

This example is from Young Lives wherein 'big qual data' are integrated within a wider panel study of childhood poverty. Over the past twenty years, Young Lives has been studying the development and wellbeing of 12,000 children in two cohorts growing up in Ethiopia, India (in United Andhra Pradesh), Peru and Vietnam through mixed-methods longitudinal research. The aim is to improve understanding of the way poverty influences children's life chances from birth to early adulthood and to explain how policies and programmes might help.

When the study was set up in 2001, it was impossible to anticipate at the outset all the lines of enquiry that would subsequently emerge as important, or the need to introduce new methods and data formats. What initially began as a household panel survey eventually expanded to include qualitative longitudinal research, qualitative sub-studies, school-based surveys, and most recently, a COVID-19 phone-based survey, all drawing on the original sample of children. So far, there have been five rounds of household surveys with the full sample and four waves of qualitative longitudinal data collection with a nested sample of over 200 children, in addition to other data exercises.

Linking numerical data produced by the survey for variable-based analysis with narrative data generated by the qualitative research for person-centred and thematic analysis poses both practical and conceptual challenges. Data governance is therefore a central feature of the study and data managers hold key roles within the team; they develop straightforward systems to allow researchers to link the qualitative, school survey and household survey data and to track all the data for a particular individual across the different waves of interviews and research components.

Each of the 12,000 children was assigned a unique ID at the beginning of the study, as were the 80 locations they were living in at the time. This facilitates drawing together large and ever-increasing volumes of very different types of data into a single corpus. For example, when the qualitative longitudinal research component was introduced in 2007, the data manager in Oxford, UK worked closely with the researchers to develop a protocol for naming data files so that these can be linked within and across data waves and countries, and to the survey data, as the example below shows.

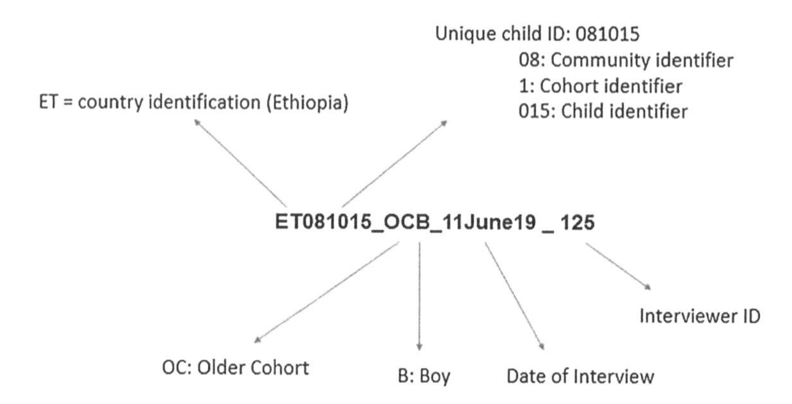

Qualitative researchers use the child IDs (as in the example, ET081015) to extract information from the surveys about individual children and their households so they can update longitudinal profiles of the children they plan to return to interview.

Moreover, it is possible to draw conceptual linkages between survey data and qualitative data to support mixed-methods analysis. Qualitative data are coded using the ATLAS-ti computer programme according to a shared coding framework, and many of the codes can be mapped against the variable names contained in the surveys, making it possible to explore a research question with data from both data sets, e.g. based on codes and variables related to 'time-use' or 'education' or 'wellbeing', etc.

Increasingly, researchers within the project are seeking ways to match Young Lives survey data to other, external, data sources, such as to administrative data and to weather data. These exciting developments promise increased analytic scale and power, at the same time, requiring that the project's systems for managing and governing data attend to new challenges in balancing consistency with flexibility, and ethical commitments to the study participants.

Case Study 4.4 QSA in the Sexuality and Abortion Stigma Study

Carrie Purcell and Karen Maxwell

The Sexuality and Abortion Stigma Study (hereafter SASS) brought together 11 qualitative data sets relating to abortion in the UK generated over a 10-year period (www.sassproject.org.uk). The study's original aim was to explore the feasibility of using qualitative secondary analysis to synthesise knowledge on manifestations of abortion stigma specifically in the UK's different jurisdictions, given that much of what was already known was US-focused. We undertook this as an interdisciplinary team with backgrounds in sociology, public health, social policy, and clinical practice. Part of our intention was to gain understanding of the resources required to conduct effective QSA versus the potential benefit from any knowledge gained. We were also keen to use QSA given the potential 'sensitivities' around the topic of abortion, and the benefit of re-examining existing data in that respect (although, as we have also explored, such presupposed sensitivities are problematic—see Kneale et al. 2019).

Key considerations in constructing our data set were: (1) whether, and to what extent, to attempt integration of diverse data with regards to type and context of production (where these factors might limit how meaningful comparison could be); and (2) the practicalities of accessing, collating, and reconciling data from different sources into a coherent whole.

The 11 original data sets focused on experiences of women who had undergone abortion and health professionals involved in provision, as well as attitudes from wider society. These came from one-to-one interviews, face-to-face and online (asynchronous) focus groups. At least one member of the project team was involved in each of the original studies, which facilitated access—as none had been deposited in a publicly available archive—as well as understanding of the original context of data production. Data sharing was overseen by data sharing agreements between our institution and those of the original studies. While we had conducted scoping work at the funding application stage to ensure consent for re-use of data was in place, the practicalities of agreeing and signing off on institutional data sharing agreements and facilitating secure transfer of data sets (including relevant meta-data) was time-consuming and took a number of weeks at the start of the study. Once compiled, the 11 studies together comprised 226 data sources and over 400 participants.

In order to explore what analyses might be possible with this vast data set, we first collated and categorised the data in NVivo, which supported comparison across data types. This review stage quickly found that comparing women's accounts of abortion would be relatively easy. However, how to integrate attitudinal data required more careful consideration. Moreover, while we were interested in cross-jurisdictional comparison (primarily due to different abortion laws and restrictions in different parts of the UK), it became evident that differences in data production could make meaningful comparison challenging, particularly with regard to some of the specificities on which the original studies focused (young people's attitudes to abortion in Scotland, attitudes to abortion in the workplace in Northern Ireland). As such, we took the decision to prioritise the experiential data as a sub-sample, with a view to applying any new learning to the attitudinal data at a later point.

Initial high-level analysis of women's and providers' accounts highlighted commonalities in the language used which shed new light on the issue of stigma (see Purcell et al., 2020; Maxwell et al., 2021). This suggested the data could be re-examined for ways in which dominant abortion narratives were challenged, and this guided the direction of our subsequent in-depth analysis.

References

Kneale, D., French, R., Spandler, H., Young, I., Purcell, C., Boden, Z., Brown, S. D., Callwood, D., Carr, S., Dymock, A., Eastham, R., Gabb, J., Henley, J., Jones, C., McDermott, E., Mkhwanazi, N., Ravenhill, J., Reavey, P., Scott, R., Smith, C., Smith, M., Thomas, J., & Tingay, K. (2019). Conducting sexualities research: An outline of emergent issues and case studies from ten Wellcome-funded projects. *Wellcome Open Research, 4*, 137. https://doi.org/10.12688/wellcomeopenres.15283.1

Purcell, C., Maxwell, K., Bloomer, F., Rowlands, S., & Hoggart, L. (2020). Toward normalising abortion: Findings from a qualitative secondary analysis study. *Culture, Health & Sexuality, 22*(12), 1349–1364. https://doi.org/10.1080/13691058.2019.1679395

Maxwell, K. J., Hoggart, L., Bloomer, F., Rowlands, S., & Purcell, C. (2021). Normalising abortion: What role can health professionals play? *BMJ Sexual & Reproductive Health, 47*, 32–36.

SASS project website. https://www.sassproject.org.uk/

Geraghty, R., & Gray, J. (2017). Family rhythms: Re-Visioning family change in Ireland using qualitative archived data from growing up in Ireland and life histories and social change. *Irish Journal of Sociology, 25*(2), 207–213.

References

Corti, L. (2000). Progress and problems of preserving and providing access to qualitative data for social research—The international picture of an emerging culture. *Forum Qualitative Sozialforschung Forum: Qualitative Social Research, 1*(3). https://doi.org/10.17169/fqs-1.3.1019

Corti, L. (2007). Re-using archived qualitative data—Where, how, why? *Archival Science, 7*, 37–54.

Crow, G., & Powell, A. (2010, July 5–8). What is missing data in qualitative research? In *NCRM research methods festival 2010*. St. Catherine's College. https://eprints.ncrm.ac.uk/id/eprint/1494

Dodds, C., Keogh, P., Bourne, A., McDaid, L., Squire, C., Weatherburn, P., & Young, I. (2021). The long and winding road: Archiving and re-using qualitative data from 12 research projects spanning 16 years. *Sociological Research Online, 26*(2), 269–287. https://doi.org/10.1177/1360780420924044

Edwards, R., & Cabellero, C. (2010). Lone mothers of mixed racial and ethnic children: Then and now. *Runnymede Bulletin, Summer Issue*, 16–17. The Runnymede Trust | Lone Mothers of Mixed Racial and Ethnic Children: Then and Now.

Faniel, I., & Yakel, E. (2011). Significant properties as contextual metadata. *Journal of Library Metadata, 11*(3–4), 155–165.

Hammersley, M. (1997). Qualitative data archiving: Some reflections on its prospects and problems. *Sociology, 31*, 131–142.

Hammersley, M. (2010). Can we re-use qualitative data via secondary analysis? *Sociological Research Online, 15*(1), Article 5.

Heaton, J. (2019). Secondary analysis of qualitative data. In P. Atkinson, S. Delamont, A. Cernat, J. W. Sakshaug, & R. A. Williams (Eds.), *Sage research methods foundations*. SAGE.

Kuula, A. (2000). Making qualitative data fit the data documentation initiative; or vice versa? *Forum Qualitative Sozialforschung / Forum: Qualitative Social Research, 1*(3).

Mason, J. (2007). "Re-using" qualitative data: On the merits of an investigative epistemology. *Sociological Research Online, 12*(3), 1–4.

Mason, J. (2018). *Qualitative researching*. Sage.

Moore, N. (2007). (Re)Using qualitative data? *Sociological Research Online, 12*(3), 1–13.

Smioski, A. (2011). Archiving qualitative data: Infrastructure, acquisition, documentation, distribution. Experiences from WISDOM, the Austrian data archive. *Forum Qualitative Sozialforschung / Forum: Qualitative Social Research, 12*(3), Art. 18.

Tarrant, A. (2017). Getting out of the swamp? Methodological reflections on using qualitative secondary analysis to develop research design. *International Journal of Social Research Methodology, 20*(6), 599–611.

Weller, S. (2022). Fostering habits of care: Reframing qualitative data sharing policies and practices. *Qualitative Research*. https://doi.org/10.1177/14687941211061054

Wilson, S. (2014). Using secondary analysis to maintain a critically reflexive approach to qualitative research. *Sociological Research Online, 19*(3), 53–64.

Part III

Moving Between Breadth and Depth in Qualitative Analysis

'Geophysical Surveying': Recursive Surface 'Thematic' Mapping Using Data Mining Tools

5

5.1 Introduction

The previous chapter ended with the making of a new assemblage of data which may be drawn from multiple project data sets in order to form a new data set or 'corpus'. This included the use of metadata to identify projects of interest and to help you identify data that is the best fit for your research aim among available secondary sources. It also included the preparatory steps of more systematic auditing of data. This preparatory work would ideally also be informed by your plan for the next step. This chapter moves to the second step in our breadth-and-depth method which uses computational textual analysis, including tools that are sometimes called data mining, to provide a detailed 'surface map' of the new data assemblage. In this step we describe a range of computational techniques in sufficient detail to convey a basic understanding of the procedures and particularly the underlying assumptions and logics used in the process. Knowing these basics of computational textual techniques and 'text mining' should help the reader to decide how to navigate this next step.

To summarise, this chapter will cover:

- a brief overview of the metaphor 'geophysical survey'
- an introduction to a range of computational textual analysis techniques
- how these forms of big qual textual analysis differ from computer aided qualitative data analysis software (CAQDAS)
- why 'text mining' is surface mapping rather than 'depth'
- a review of the capacity knowledge and skills required for the computational techniques
- the basics of computational textual techniques and 'text mining' without the maths

As noted in Chap. 2, here 'mining' does not mean deep digging. In terms of our archaeological metaphor, this step is equivalent to the geophysical survey in which the selected area of landscape is painstakingly documented on foot—walked over

© The Author(s), under exclusive license to Springer Nature Switzerland AG 2023

S. Weller et al., *Dig Qual*, https://doi.org/10.1007/978-3-031-36324-5_5

and marked off in squares that record the co-ordinates of surface features. However, our landscape consists of an assemblage of texts, and it is computer programs that are designed for text analysis that are doing the moving back and forth across the landscape to provide the detailed scrutiny and documentation of the surface. Note that in what follows the new data set assembled in step one is sometimes referred to as the 'corpus' because this is the term often used in linguistics for assemblages of natural language and many of the computational techniques have been developed for corpus linguistics, often in cross-pollination with computer science. In the breadth-and-depth method, however, the surface mapping that the computational tools provide is not an end in itself. Just as the geophysical survey might be used to inform the archaeologists' next steps of digging test pits to further narrow down where to dig deep, our surface map is a step towards choosing where and when to begin qualitative analysis which involves a human researcher in detailed reading of specific texts (Fig. 5.1).

In our view, the surface map of step two is unlikely to ever be sufficient as an endpoint when the overarching aim is to maximise the possibilities of new insights from an assemblage of qualitative research projects. Many of the computational

Fig. 5.1 Archaeological metaphor and step two of the breadth-and-depth method. (Illustration created by Chris Shipton https://www.chrisshipton.co.uk/)

tools used in this step take a 'bag of words' approach that treats the data as if nothing is known about the speaker or the context of the speaking (see Wiedemann, 2013 for more detailed discussion of the history of the approach in Natural Language Processing). To qualitative researchers who have carefully archived their project's data, a bag of words approach may seem like violence but without computational textual techniques there is no possibility of systematic analysis of big qual accumulations of such data. However, to apply computational mapping without taking on board prior knowledge of the data or to stop at the mapping with no effort to connect back to the richness of the data seems to us to be an equally dismal disregard of the opportunities on offer. While our advocacy of computational textual techniques will be a sore challenge to some qualitative researchers, our insistence on a return to more conventional qualitative methods will be a disappointment to readers seeking a quick computational fix to the challenge of analysing more qualitative data than can easily be read by a human. In learning each other's crafts, both stand to gain.

Step two in the breadth-and-depth method, as with every step, is using techniques for a purpose that is shaped by the researcher's theoretically informed approach and research questions. The nature of the new secondary data set assembled in step one and the purposes it reflects should customise how computational techniques are used as a strategy for surveying the 'breadth' of the data in order to decide where to return to 'depth'. If, for example, in the previous step, the data set has been designed to purposively bring data together for comparison, then realisation of the possibilities of comparison will structure how computational text analysis is used in step two. The way text analysis programs are used will also vary according to whether the approach is theoretically framed as a process of discovery that is entirely 'from the data' or a form of hypothesis testing that involves seeking pre-identified terms with their identification driven by theory. In either case, a researcher is likely to continue to make use of the metadata attached to the source project and particular texts.

General social science research methods texts do not yet typically reference the computational techniques described in this chapter. For example, the fifth edition of Bryman's popular text *Social Research Methods* (Bryman, 2016) makes no reference to computer programs when discussing content analysis or big data, although the section on qualitative research has a chapter on the analysis program, NVivo. Recent specialist texts on content analysis (e.g. Krippendorff, 2019) are more likely than general social research methods books to include social science-driven discussion of data mining. Mills' text discussed in Chap. 2 and Weidemann's monograph are perhaps examples in the vanguard of what may become a genre of texts focused specifically on big qual (Mills, 2019; Wiedemann, 2016). While we share their view that techniques of data mining create new possibilities for qualitative social researchers, we are less sure of Weidemann's confidence that the gap between 'how qualitative researchers perceive their object of research' and 'what computer algorithms are able to identify is constantly narrowing' (2016, p. 22)—or its desirability where bias is often built into algorithms. However, more researchers trying out the breadth-and-depth method may encourage constructive dialogue and mutual learning between researchers who brand themselves as qualitative and computational or quantitative and encourage beneficial cross-over in techniques.

5.2 CAQDAS Software Not the Only Tool in the Box

As noted in Chap. 1, use of CAQDAS has become a standard aspect of doing qualitative research. For social science researchers in some academic contexts within the English-speaking world, NVivo enjoyed a period as the market leader. The term 'text mining' would not normally be applied to CAQDAS. Rather, CAQDAS software builds from the conventional manual processes of qualitative data analysis that always start with listening to, watching, or reading individual data records (e.g. an interview recording or transcript, a fieldwork video or note, a diary entry, a significant document) in order to systematically pick out meanings, concepts, themes, people, events, narratives, processes, and the like that are interpreted by the analyst as significant and worthy of coding in the data. Human analysts often simultaneously develop different levels and types of codes ranging across descriptive and analytical categories that already express theorisation of data. The fundamental facility offered by the CAQDAS software is ease of coding and of using codes to assemble data for further analysis. The software facilitates attaching codes to segments of text, whatever these signify to the analyst, and speeds up the process of extracting specific coded segments across all coded data records for further scrutiny while always retaining a direct route back to the original context.

CAQDAS software typically also facilitates more automated data exploration such as searching for words or phrases and automatically coding a predetermined amount of surrounding data with that word or phrase. However, the basic design assumption remains that the software will be used in the context of human familiarity with the data through repeated listening, watching, or reading the data. In contrast, the data mining tools being discussed here have been developed by scholars of natural language, information retrieval, and computational science for automated forms of analysis designed with no assumption of reading by a human researcher. Rather the analysis is built through computational steps such as counting the number of unique words and their frequency in each document, calculating their relative frequency against all documents, and the recursive comparative scrutiny of patterns of co-location and frequency of occurrence between documents. While the amount of text tackled by CAQDAS software is restricted by the demands on the researchers as analysts, data mining can be applied to almost any amount of text and uses procedures far removed from human-researcher concerns with the context and meaning of words.

5.3 Surface Sifting, Not 'Mining' as Depth

From the perspective of a social researcher trained in qualitative data analysis, the label of 'mining' in the term 'data mining' that is often applied to CAQDAS is a misnomer because it suggests digging deep, but 'depth' has particular meanings within qualitative research. Depth suggests exploring details of social process, nuances of meanings, and constant attention to links between micro-details and wider contexts. This is not what the computational techniques used in step two achieve. In terms of our archaeological metaphor, we are still mapping the surface. The way that computational techniques 'map' the words that make up the 'landscape' is better imagined as surface-searching than digging. If a 'mining' analogy is to be retained, then think 'surface strip mining' and not 'deep shaft mining', but, of course, in a digital environment, unlike a physical world, this 'strip mining' can occur without permanently destroying the landscape—words are sifted without permanently destroying the particular order of words that comprise the texts.

For the qualitative researcher understanding is deepened by constantly referring back to the wider context of any particular data extract. Here 'context' includes how data were elicited, as well as how a particular extract fits in a bigger story being told or reflects known circumstances revealed outside of its immediate telling. Detailed consideration of how data are multiply related to and framed by context grounds the process of constant comparison across apparently similar data extracts to tease out and theorise difference, nuance, levels, and layers of meaning. Rather than this kind of 'depth', the outputs of the computational techniques of data mining are better described as recursive 'topic' or 'thematic' surface mapping. Program developers use different labels for the discovered meaningful words that their program produces as the best probability of representing a semantically coherent word collection in the data, including 'topic', 'theme', and 'concept'. Whatever the term used, the basic techniques of 'text mining' are broadly similar and seem superficial in comparison to the range of factors being weighed up by a qualitative researcher engaged in coding activities and developing a coding frame during in-depth qualitative analysis. In the breadth-and-depth method the results of data mining techniques are not being treated as the endpoint but as one step in an iterative process between breadth and depth. The results produced by data mining are the starting points for sampling short extracts of data in step three. Taken together, steps two and three are a systematic, rigorous, and considered way of moving from breadth to depth.

5.4 Capacity, Knowledge, and Skill Required

Within the breadth-and-depth method, step two may be the most challenging for some qualitative researchers who are in the relatively common situation of being unfamiliar with the computer aided techniques involved. Although PhD training in the social sciences usually includes modules in the analysis of both quantitative and qualitative data, the typical suite of courses does not yet routinely include an introduction to 'data mining' techniques, although this is changing. As with statistics, there are software packages that can be used without a thorough understanding of the computational steps involved, but this carries obvious risks of applying techniques inappropriately, and misunderstanding and misrepresenting the results. Unlike the various generations of the Statistical Package for the Social Sciences (SPSS) or its potential successor Stata, there are not yet clear lead products, multiply tried and tested by generations of social scientists. Rather there is a growing and incrementally changing range of products sometimes using variations in techniques or describing similar techniques in rather different terms. In this text, we are reluctant to name specific products because of the speed of change but there is a clear case for advocating on behalf of programs that are freeware or 'open source'. In the field of linguistics, the case for collaboratively building and refining freeware tools has been made by Laurence Anthony (2009) and much of his argument applies to social science. Open-source products are transparent and benefit from the critical engagement of user communities. Moreover, social scientists as users will have more opportunity to make rapid progress with such open-source programs as R and Python if it becomes common practice for published results to be accompanied by shared code. As Nelson (2020, p. 11) notes, such open-source programs provide users with the most flexible and reproducible approach. Both programs are well documented by online tutorials and secondary literatures, albeit often more geared to data scientists than social scientists (Jockers, 2014; Lutz, 2001; Bonzanini, 2016). They are not, of course, the only option, as other largely open-source languages have also been used to develop text analysis software, such as the Java-based MALLET (Graham et al., 2012) and Perl-based software Antconc (Anthony, 2009, 2022).

The product pool includes commercial text analysis software that claims to map topics without a steep learning curve and that produces easily read outputs such as word clouds or diagrammatic representations of more or less dominant themes; it offers to bypass the need to learn a programming language in order to get to this point. Chapter 6 provides some examples of one such program, Leximancer. Note that the computational steps involved in such products are typically described in rather broad-brush terms because it would be commercially disadvantageous to provide complete details. If users detail their steps within the program and are using publicly available data, their work could be reproduced by another researcher using that same program. However, this level of transparency is not sufficient to guarantee the integrity and rigour of the work. If conducted with no understanding of the basics of text mining, the researcher risks inappropriate use and exaggerated claims, which might then be repeated by the next no-better-informed user.

The next section offers a summary of the basic techniques. Awareness of the basics should ideally be combined with an understanding of how the techniques are being applied in the particular bespoke software being used but, because commercial products are not totally transparent, it may not be possible to fully spell this out. A sense of safety in numbers will no doubt emerge as product usage picks up and brands are more intensively discussed in the literature. As yet, there is limited work comparatively applying products to the same complex qualitative analytical tasks, although there is a significant body of work that has made a start including computational scientists addressing a wider audience (Blei, 2012) and the growing body of early adopters in the social sciences (DiMaggio, 2015).

Adopting a freeware approach need not mean going it alone or starting from scratch, since there is a legacy of a growing number of tools that others have already developed in open-source programming languages such as R and Python. In our own teaching of the breadth-and-depth method we have used Antconc, a program developed for linguistics in the language Perl, because it is free, users can learn it quickly, it offers many of the techniques for computer analysis of text that may be of interest to a qualitative researcher, and it may suffice for some purposes. However, it does not provide the form of topic modelling described below that uses 'machine learning' to discover topics in text and is often seen as the most promising aspect of 'data mining' (Wiedemann, 2016).

Being able to range across the whole gamut of possibilities requires learning an open-source programming language such as R or Python. The effort of building ease in such a language is probably no more demanding for a social scientist with no programming skills than learning SPSS was in the very early days when commands had to be learned and typed accurately without the help of prompting dropdown menus, as data processing was still done on 'mainframe' computers and submitted on punch cards typed out on a card writing machine. However, the effort is sufficient to make this difficult to justify unless the skill is going to be actively used or you simply enjoy a challenge. The learning curve makes this less attractive in mid-late career. An alternative route is to work in teams with colleagues who are already practitioners. However, computationally skilled users of the relevant techniques are most often found in disciplines beyond those inhabited by qualitative researchers, such as computer science. As in all interdisciplinary collaborations, translation work will then be required, as the basic principles of qualitative research and working in 'qualitatively driven ways' may be as alien to some practitioners with ease in programming as reducing text to 'bags of words' is to some qualitative researchers. Recruiting a skilled practitioner as a technical assistant who has no interest in or commitment to the research questions seems likely to risk mutual frustration. On the one hand, ease of translation is likely to be advanced as successful collaborations multiply; on the other, the need for collaboration is likely to reduce as more social scientists become programmers.

Even for the mathematically and computer literate social scientist who can very quickly become an expert user of any chosen software, this step needs time. The idea that computing makes for fast work needs to be tempered with several caveats. Text typically requires preparation to be in the right form for computer text analysis or 'data mining' involved, whether saving all files as simple text or the sequence of more complicated transformations necessary for many programs. As with all academic research, being able to give an account of and document what you are doing and why is essential from the outset, and this also inevitably takes time. Step two is likely to be an iterative trial-and-error, try-and-try-again process that must be carefully documented in order to progress your own learning and stand up to scrutiny. The whole process might be very slow and painstaking.

5.5 Basics of Computer Text Analysis and 'Text Mining'

What is a word? A sentence? Understanding how a word has been defined for computation is important in keeping a firm grip on the enormous difference between machine 'reading' and human reading. Sometimes the term 'token' rather than 'word' is used to denote the transformation into machine-readable units of analysis. The term 'vector' indicates the additional step of turning strings of 'tokens' into a string of numbers.

With respect to words in the English language, a computer programming instruction to break a text into 'tokens' might start by defining a word as any continuous string of any of the letters from a to z followed by a blank space. This would miss words with apostrophes, hyphens, and followed by punctuation, so extra bits of programming are required to enable their recognition as words. In terms of an interest in identifying words that are signifiers of meanings or topics, it may not be desirable to treat the plural and singular form of a noun or different parts of the same verb as different words. The terms lemmatisation and stemming are used for the process of grouping together different forms of the same content-word so that they can be counted as one word. Similarly, it is not ideal to treat words that are identical except beginning with an upper-case or lower-case letter as different words. Existing web-based literatures around computer aided textual analysis using particular programs offer many examples to follow in dealing with these steps in pre-processing text ready for computational analysis (e.g. Python Natural Language Toolkit, n.d.; Porter Stemmer, n.d.; Porter 2, n.d.). For some documents, like notes or records produced on typewriters or by hand, there are additional challenges of digitization. For some texts further processing might involve correcting common spelling mistakes so that misspellings are not treated as different words.

Just as rules can be constructed to define 'words' in order to make the outcome the basic unit of analysis for computation, programmers have defined a sentence or a paragraph in order to make them units of analysis. Some text pre-processing programs offer this as an option. As with 'words', the outcome is only as good as the rules permit; for example, the variable uses of punctuation by authors and transcribers are likely to modify what is recognized.

When the intention is to scrutinize words to see what they indicate about the pattern of topics, very frequent and very rare words are sometimes discarded as part of the process of turning text into a bag of words for computation. For example, Grimmer and King describe what they see as the standard text pre-processing before conducting their analysis: "We transform to lower case, remove punctuation, replace words with their stems, and drop words appearing in fewer than 1% or more than 99% of documents" (Grimmer & Gary, 2011, p. 2644). The authors are treating very rare words as too idiosyncratic to be 'topics', and very frequent words typically appear in the standard lists of 'stop words' that are routinely removed.

5.5.1 Throwing Away 'Stop Words'

In forms of text analysis that are motivated by an interest in content meaning rather than sentence structure, many words are regarded as 'stop words', unimportant clutter that would be best removed. This includes words that are used particularly frequently like 'a', 'an', 'and', 'it', and 'the'. A bit of googling will quickly reveal suggested lists of stop words, and text analysis programs typically include the option of using such a list (e.g. Porter 2, n.d.). Freeware programmes for computational textual analysis often suggest stop word lists that are intended to remove all but 'content words', words that carry content or meaning, referred to as 'meaningful words' in Case Study 5.1 by Seale and colleagues.

Note that many, perhaps most, of the words in the example sentence 'and stuff like that so it made me angry a bit' would have been knocked out by a stop word list. This would certainly be the case for 'a', 'and', 'so', and 'that'. A researcher who is determined to eliminate all words that do not convey specific content might go further to include the vague terms such as 'bit' and 'stuff' and the ambiguous term 'like' which is used as a form of punctuation by some English-language speakers, as well as to convey a positive preference. This would mean that only the term 'angry' would survive.

5.5.2 Word Counting: Comparing Frequencies and 'Keyness' of Words

The frequency distribution of words plays a key part in seeking to discover topics through 'text mining' but there are also more basic procedures based on frequencies that seek to identify words that seem to be more important than others, 'keywords'. Assuming 'stop words' have been removed and a research design that includes comparison, the relative frequency of words in particular texts or sets of texts can start to become interesting by suggesting differential salience or 'keyness' of meaningful words between the comparators. Our interest in a hypothesised gender convergence in vocabularies and practices of care and intimacy over time could be partially advanced by using frequencies—for example, comparing the words of older men versus older women and doing so in comparison to the words of young men versus young women. In the relevant version of our corpus, interview transcripts from projects in the Timescapes Data Archive with the words of the interviewer removed are organised into subsets by age cohort and gender. Once stop words are applied, even just looking at relative word frequencies (i.e. looking at the rates or proportions of their use in order to control for text size) may suggest interesting differences in the 'naturally occurring' most frequent words. We tried the strategy of looking comparatively at what words were most frequent and, alternatively, looking specifically at the relative frequency of words that we, as analysts, recognise as being about care and intimacy. The latter approach is common in content analysis, which often proceeds from a researcher-prepared dictionary of words of interest to look comparatively at their relative frequency across texts.

5.5.3 The Framing of Words

Comparing frequencies or using more sophisticated algorithms to score words and suggest keywords that may indicate meaningful topics must proceed with mindfulness of how the data have been structured by the research process. This is an obvious issue when conducting secondary analysis of archived, social researcher-generated qualitative data, drawing data across different projects. It is imperative to consider how the relative frequency of particular words has been influenced by the researcher rather than reflecting their 'natural' salience for particular groups, and ignoring the fact that research participants' words in interviews are influenced by the questions asked. Removing the researchers' words from the data set ensures only giving weight to the research participants' words, but it does not, of course, erase the consequences of participants responding to particular questions.

In the case of the Timescapes Data Archive example, in one project, questions reflect the primary focus on friends and siblings; others tailor questions to their concerns with parenting and/or grandparenting. It would not be surprising, therefore, to find differences in the frequency of words concerning these particular types of personal and familial relationships. The way in which comparison is being used then becomes important. Older men and older women are drawn from projects that focus on parenting and grandparenting while young men and women are drawn from projects that focus on siblings and friends. Arguably, therefore, in this case comparison by gender within cohort has more legitimacy than a comparison by age within gender, and this has to be acknowledged when presenting data.

5.5.4 RAKE (Rapid Automatic Keyword Extraction)

This is a well-used and documented freeware that offers a more sophisticated set of procedures for identifying keywords than relative frequencies using a combination of word counting and algorithms scrutinising the colocation of words, that is, their proximity operationalised as the words within a set span before and/or after each potential keyword. The pre-processing steps deal with unwanted punctuation and lemmatisation and use a stop word list. The algorithms then use the colocation of words to split the words into phrases of content words. In the spoken sentence 'I get angry with her because before she let me borrow her shoes and I messed them up a bit and she got angry about that and wouldn't let me' the stop words would include: I, with, her, because, before, she, me, her, and, them, up, a, about, that. Because of their vagueness the verbs get, got, wouldn't (both would and not if pre-processing of apostrophes had changed 'wouldn't' to 'would' and 'not') would also be treated as stop words. This would only leave six content words: angry, let, borrow, shoes, mess, bit. If only immediately collocated words are counted as phrases the six words would also each be phrases since they are not immediately next to another content word, but if the algorithm allows for one stop word in between content words then the phrases are reduced to four: 'angry', which appears twice, 'let borrow shoes', 'mess', and 'let'. Algorithms are looking for recurring patterns and so if words or

Table 5.1 Matrix of words in example sentence

	Angry	Let	Borrow	Shoes	Mess	Bit
Angry	2	0	0	0	0	0
Let	0	2	1	1	0	0
Borrow	0	1	1	1	0	0
Shoes	0	1	1	1	0	0
Mess	0	0	0	0	1	0
Bit	0	0	0	0	0	1

phrases appear in the same order several times then the content words will be recognised as a phrase even with several stop words in between.

Having broken the text down in this way, a matrix showing the frequency of each content word and the number of co-occurrences it has in phrases with each other content word would be constructed. A matrix of the example sentence is given in Table 5.1. This simple matrix is the basis of calculating a score for every word—the number of times it co-occurs with another content word in a phrase, divided by the frequency of its every occurrence. Phrases are also given a score by adding up the scores of each word in the phrase. This formula is used to identify the top words and phrases as the 'keywords' and enable comparative comparison of the 'keywords' in different sets of documents.

5.5.5 Other Approaches to Keyness

While 'keyword' is often used to indicate a word that may represent a particular topic, it is sometimes given a more specific definition that includes reference to the relative salience or significance of a topic. This usage adds 'keyness' to 'keyword'. For example, English-language scholar Scott (1997) referred to words which occur with unusual frequency or infrequency in a text in comparison to the frequency of the same word in a relevant 'reference corpus'. An English-language local-dialect specialist, for example, might be interested in comparing vocabulary used in a particular village with a very large data set of naturally occurring language across the whole country, but the relevant 'reference corpus' depends on research questions and research design. The basic computational procedure is looking for significant differences between the proportion of each term comprising all words in one text or group of texts versus their proportion in another relevant comparator text or group of texts. Nelson's use of difference of proportions is described in Case Study 5.2. Software for text analysis like Wordsmith that Scott was using, or the freeware Antconc, allows you to quickly produce a rank of 'keyness' with a procedure for this specific purpose. The results show a list of words with exceptional frequency (or infrequency) in one file or set of files in comparison to your reference corpus and provide the statistical significance. How Seale, Ziebland, and Charteris-Black used keyness to identify differences in men's and women's ways of speaking about their health is described in Case Study 5.1. They used a sample of men's and women's narratives of health and illness selected from a large database and matched in terms

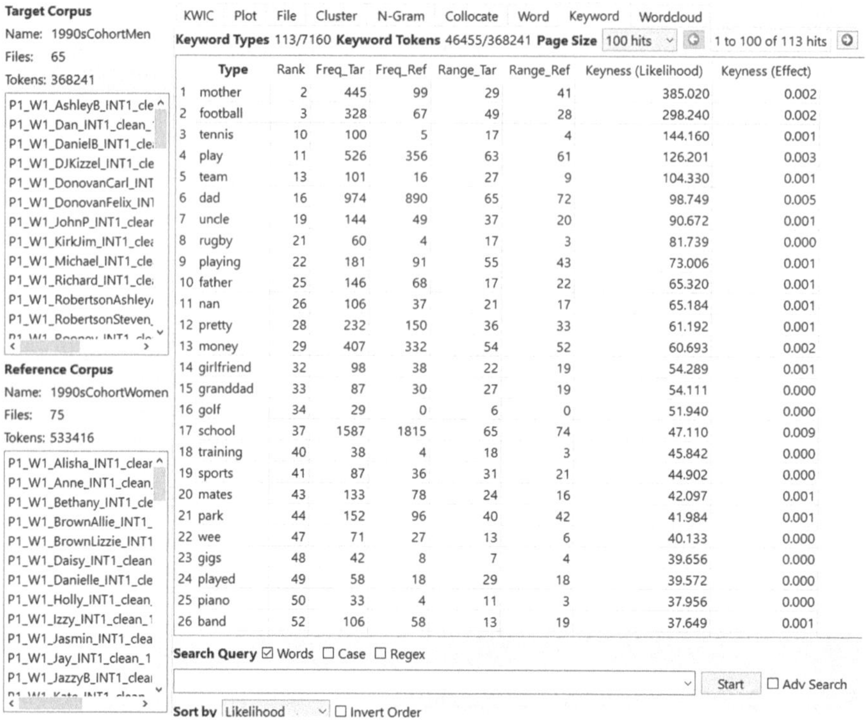

Target Corpus
Name: 1990sCohortMen
Files: 65
Tokens: 368241

P1_W1_AshleyB_INT1_cle
P1_W1_Dan_INT1_clean_
P1_W1_DanielB_INT1_cle
P1_W1_DJKizzel_INT1_cle
P1_W1_DonovanCarl_INT
P1_W1_DonovanFelix_INT
P1_W1_JohnP_INT1_clear
P1_W1_KirkJim_INT1_cle
P1_W1_Michael_INT1_cle
P1_W1_Richard_INT1_cle
P1_W1_RobertsonAshley
P1_W1_RobertsonSteven

Reference Corpus
Name: 1990sCohortWomen
Files: 75
Tokens: 533416

P1_W1_Alisha_INT1_clear
P1_W1_Anne_INT1_clean
P1_W1_Bethany_INT1_cle
P1_W1_BrownAllie_INT1_
P1_W1_BrownLizzie_INT1
P1_W1_Daisy_INT1_clean
P1_W1_Danielle_INT1_cle
P1_W1_Holly_INT1_clean
P1_W1_Izzy_INT1_clean_1
P1_W1_Jasmin_INT1_clea
P1_W1_Jay_INT1_clean_1
P1_W1_JazzyB_INT1_clea

KWIC Plot File Cluster N-Gram Collocate Word Keyword Wordcloud

Keyword Types 113/7160 **Keyword Tokens** 46455/368241 **Page Size** 100 hits 1 to 100 of 113 hits

	Type	Rank	Freq_Tar	Freq_Ref	Range_Tar	Range_Ref	Keyness (Likelihood)	Keyness (Effect)
1	mother	2	445	99	29	41	385.020	0.002
2	football	3	328	67	49	28	298.240	0.002
3	tennis	10	100	5	17	4	144.160	0.001
4	play	11	526	356	63	61	126.201	0.003
5	team	13	101	16	27	9	104.330	0.001
6	dad	16	974	890	65	72	98.749	0.005
7	uncle	19	144	49	37	20	90.672	0.001
8	rugby	21	60	4	17	3	81.739	0.000
9	playing	22	181	91	55	43	73.006	0.001
10	father	25	146	68	17	22	65.320	0.001
11	nan	26	106	37	21	17	65.184	0.001
12	pretty	28	232	150	36	33	61.192	0.001
13	money	29	407	332	54	52	60.693	0.002
14	girlfriend	32	98	38	22	19	54.289	0.001
15	granddad	33	87	30	27	19	54.111	0.000
16	golf	34	29	0	6	0	51.940	0.000
17	school	37	1587	1815	65	74	47.110	0.009
18	training	40	38	4	18	3	45.842	0.000
19	sports	41	87	36	31	21	44.902	0.000
20	mates	43	133	78	24	16	42.097	0.001
21	park	44	152	96	40	42	41.984	0.001
22	wee	47	71	27	13	6	40.133	0.000
23	gigs	48	42	8	7	4	39.656	0.000
24	played	49	58	18	29	18	39.572	0.000
25	piano	50	33	4	11	3	37.956	0.000
26	band	52	106	58	13	19	37.649	0.001

Search Query ☑ Words ☐ Case ☐ Regex

Start ☐ Adv Search

Sort by Likelihood ☐ Invert Order

Fig. 5.2 'Keyness': Young men's frequent words in comparison to young women (see warning above about 'mother')

of high and low social class profiles with the help of the attached metadata (2008, 2010).

The tables in Figs. 5.2 and 5.3 illustrate the program Antconc keyness function applied to the youngest cohort of men and women assembled from Timescapes data and signal the further work required to interpret the output (note that pseudonyms are used in all examples). Figure 5.2 shows the words used by young men with unusual frequency in comparison to young women and Fig. 5.3 shows the words used by young women in unusual frequency in comparison to young men. The tables would be quite different without stop words since young men and women use some 'meaningless' words, such as 'uh', with different frequency. The words of the interviewer are not included in the data with the intention that only the words of the interviewee should count. But this pre-processing step is imperfectly executed. Further scrutiny of 'mother', apparently used with unusual frequency by young men, demonstrates that it was not in fact a word usually used by them but inserted by the transcriber to indicate the interviewee's mother was present or speaking. The apparently comparatively higher usage of the word 'friend' among young women than young men does reflect a difference but not quite the words of the research participants because the process of anonymisation often substitutes the word 'friend'

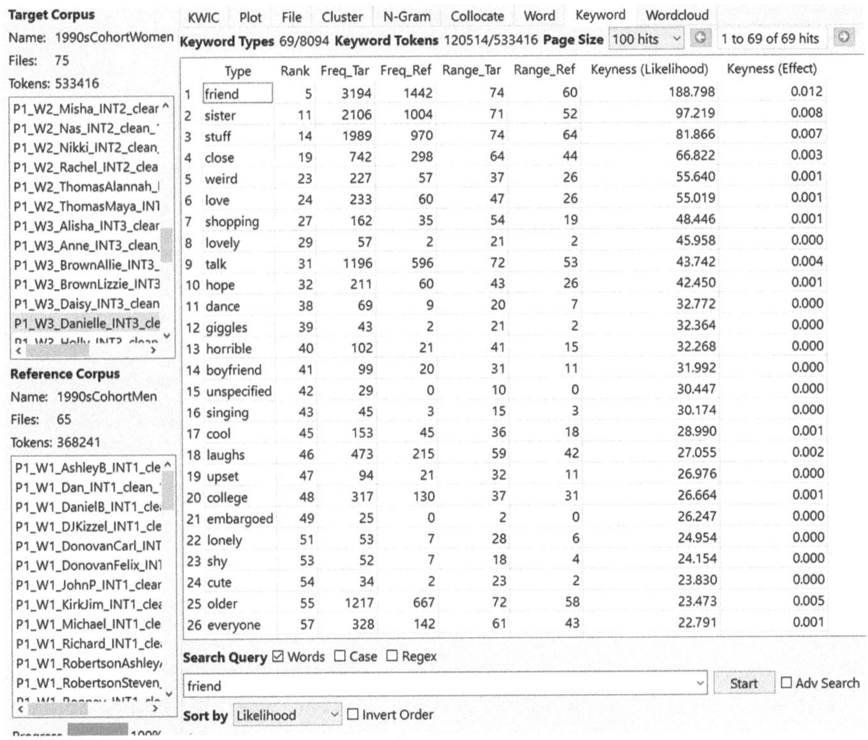

Target Corpus
Name: 1990sCohortWomen
Files: 75
Tokens: 533416

P1_W2_Misha_INT2_clear
P1_W2_Nas_INT2_clean_
P1_W2_Nikki_INT2_clean.
P1_W2_Rachel_INT2_clea
P1_W2_ThomasAlannah_I
P1_W2_ThomasMaya_IN1
P1_W3_Alisha_INT3_clear
P1_W3_Anne_INT3_clean.
P1_W3_BrownAllie_INT3_
P1_W3_BrownLizzie_INT3
P1_W3_Daisy_INT3_clean
P1_W3_Danielle_INT3_cle

Reference Corpus
Name: 1990sCohortMen
Files: 65
Tokens: 368241

P1_W1_AshleyB_INT1_cle
P1_W1_Dan_INT1_clean_
P1_W1_DanielB_INT1_cle
P1_W1_DJKizzel_INT1_cle
P1_W1_DonovanCarl_INT
P1_W1_DonovanFelix_IN1
P1_W1_JohnP_INT1_clear
P1_W1_KirkJim_INT1_cle
P1_W1_Michael_INT1_cle
P1_W1_Richard_INT1_cle
P1_W1_RobertsonAshley
P1_W1_RobertsonSteven.

KWIC Plot File Cluster N-Gram Collocate Word Keyword Wordcloud

Keyword Types 69/8094 **Keyword Tokens** 120514/533416 **Page Size** 100 hits 1 to 69 of 69 hits

	Type	Rank	Freq_Tar	Freq_Ref	Range_Tar	Range_Ref	Keyness (Likelihood)	Keyness (Effect)
1	friend	5	3194	1442	74	60	188.798	0.012
2	sister	11	2106	1004	71	52	97.219	0.008
3	stuff	14	1989	970	74	64	81.866	0.007
4	close	19	742	298	64	44	66.822	0.003
5	weird	23	227	57	37	26	55.640	0.001
6	love	24	233	60	47	26	55.019	0.001
7	shopping	27	162	35	54	19	48.446	0.001
8	lovely	29	57	2	21	2	45.958	0.000
9	talk	31	1196	596	72	53	43.742	0.004
10	hope	32	211	60	43	26	42.450	0.001
11	dance	38	69	9	20	7	32.772	0.000
12	giggles	39	43	2	21	2	32.364	0.000
13	horrible	40	102	21	41	15	32.268	0.000
14	boyfriend	41	99	20	31	11	31.992	0.000
15	unspecified	42	29	0	10	0	30.447	0.000
16	singing	43	45	3	15	3	30.174	0.000
17	cool	45	153	45	36	18	28.990	0.001
18	laughs	46	473	215	59	42	27.055	0.002
19	upset	47	94	21	32	11	26.976	0.000
20	college	48	317	130	37	31	26.664	0.001
21	embargoed	49	25	0	2	0	26.247	0.000
22	lonely	51	53	7	28	6	24.954	0.000
23	shy	53	52	7	18	4	24.154	0.000
24	cute	54	34	2	23	2	23.830	0.000
25	older	55	1217	667	72	58	23.473	0.005
26	everyone	57	328	142	61	43	22.791	0.001

Search Query ☑ Words ☐ Case ☐ Regex
friend Start ☐ Adv Search
Sort by Likelihood ☐ Invert Order

Fig. 5.3 'Keyness': Young women's frequent words in comparison to young men's

each time a personal name is used. The process of interpretation is discussed further in the next chapter.

Researchers have also mixed and switched between measures of frequency of all 'content words' and frequencies found in searches for researcher-selected words of significance for the research purposes. For example, as well as using the keyness function of the text analysis program, Seale and Charteris-Black (2008, 2010) searched for words conveying emotions, using a pre-developed system for tagging words belonging to particular categories of meaning, the 'UCREL Semantic Analysis System' (http://ucrel.lancs.ac.uk/usas/) developed at the University of Lancaster, UK.

5.5.6 Clusters of Words as 'Topics'

A fairly typical definition of a 'topic' in the text-mining approach is offered in a tutorial on 'topic modelling' using the program MALLET—'a cluster of words that frequently occur together'. However, there can be different human understandings and various computational instructions for identifying 'occurring together'. As the RAKE example has already shown, word order can be used in text analysis software to use 'content words' that are neighbours or located in proximity to each other in

Fig. 5.4 KWIC: keyword (angry) in context

order to suggest the importance of phrases and to identify 'keywords' or as Scott put it 'key key words' that occur across many texts (Scott, 1997, p. 233). Analysts of natural language have a longstanding interest in learning from word order, the concordance of words, that is, paying attention to the word that typically appears before and after a particular word, also referred to as KWIC—keyword in context. The illustration in Fig. 5.4 shows the output of a KWIC search on 'angry' using the program Antconc. In the example, the outputs are ordered by frequency attending to the three words to the right showing that 'with me' and 'with each other' were the most common combinations following 'angry' in the 140 files used. It would have been possible instead to have focused on one or more words to the left.

Using the same package and data, the illustration in Fig. 5.5 is the output ordered by statistical measures of frequency of collocation between 'angry' and another word in this collection of documents.

Finally, the last example (Fig. 5.6) shows the output using the same data to search for clusters of two to six words including 'angry' ranked by the most frequent word to follow 'angry' in such clusters.

As programs like Antconc enable immediate moving back and forth between a 'keyword' and the text it comes from, such procedures can help to quickly determine whether apparently important keywords represent anticipated topics. However, some increasingly popular computational approaches to identifying 'topics' do not use word order but treat it as completely irrelevant.

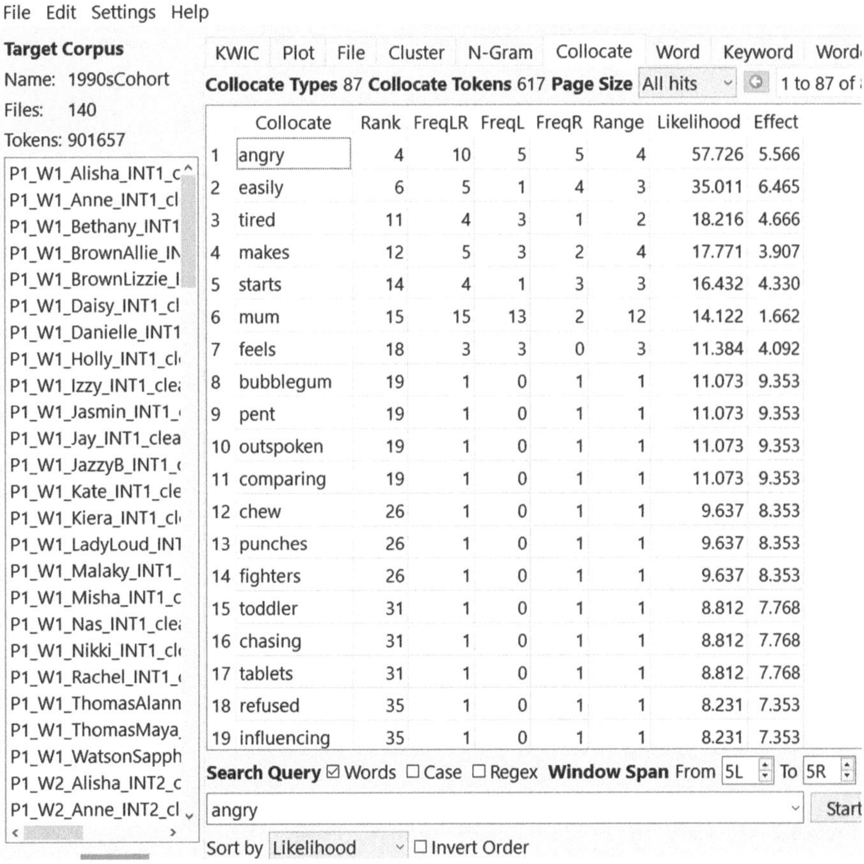

File Edit Settings Help

Target Corpus
Name: 1990sCohort
Files: 140
Tokens: 901657

P1_W1_Alisha_INT1_c
P1_W1_Anne_INT1_cl
P1_W1_Bethany_INT1
P1_W1_BrownAllie_IN
P1_W1_BrownLizzie_I
P1_W1_Daisy_INT1_cl
P1_W1_Danielle_INT1
P1_W1_Holly_INT1_cl
P1_W1_Izzy_INT1_cle
P1_W1_Jasmin_INT1_
P1_W1_Jay_INT1_clea
P1_W1_JazzyB_INT1_
P1_W1_Kate_INT1_cle
P1_W1_Kiera_INT1_cl
P1_W1_LadyLoud_IN1
P1_W1_Malaky_INT1_
P1_W1_Misha_INT1_c
P1_W1_Nas_INT1_cle
P1_W1_Nikki_INT1_cl
P1_W1_Rachel_INT1_
P1_W1_ThomasAlann
P1_W1_ThomasMaya
P1_W1_WatsonSapph
P1_W2_Alisha_INT2_c
P1_W2_Anne_INT2_cl

KWIC Plot File Cluster N-Gram Collocate Word Keyword Word

Collocate Types 87 **Collocate Tokens** 617 **Page Size** All hits 1 to 87 of

	Collocate	Rank	FreqLR	FreqL	FreqR	Range	Likelihood	Effect
1	angry	4	10	5	5	4	57.726	5.566
2	easily	6	5	1	4	3	35.011	6.465
3	tired	11	4	3	1	2	18.216	4.666
4	makes	12	5	3	2	4	17.771	3.907
5	starts	14	4	1	3	3	16.432	4.330
6	mum	15	15	13	2	12	14.122	1.662
7	feels	18	3	3	0	3	11.384	4.092
8	bubblegum	19	1	0	1	1	11.073	9.353
9	pent	19	1	0	1	1	11.073	9.353
10	outspoken	19	1	0	1	1	11.073	9.353
11	comparing	19	1	0	1	1	11.073	9.353
12	chew	26	1	0	1	1	9.637	8.353
13	punches	26	1	0	1	1	9.637	8.353
14	fighters	26	1	0	1	1	9.637	8.353
15	toddler	31	1	0	1	1	8.812	7.768
16	chasing	31	1	0	1	1	8.812	7.768
17	tablets	31	1	0	1	1	8.812	7.768
18	refused	35	1	0	1	1	8.231	7.353
19	influencing	35	1	0	1	1	8.231	7.353

Search Query ☑ Words ☐ Case ☐ Regex **Window Span** From 5L To 5R
angry Start
Sort by Likelihood ☐ Invert Order

Fig. 5.5 Collocates of 'angry'

5.5.7 The Document-Term Matrix and 'Bag of Words' Approach

At the base of much computer textual analysis are transformations that reduce a set of text documents into what is often called a document-term matrix. This involves identifying each unique word in the entire corpus and then creating a record for each document which shows whether each unique word is present or absent and, if present, the frequency with which it appears. The example in Table 5.2 is a fragment of a possible document-term matrix. The three rows represent three documents, in this case transcripts of interviews with three research participants from one project in our Timescapes multi-project corpus. The columns present an artificially truncated set of terms, only six of the over 5000 words that appear in this project. They are the content words that appeared in the example of the young woman's sentence given above, here represented as transcript 1. The example shows the frequency of these terms in this and two other young women's transcripts. Our reordering of the words

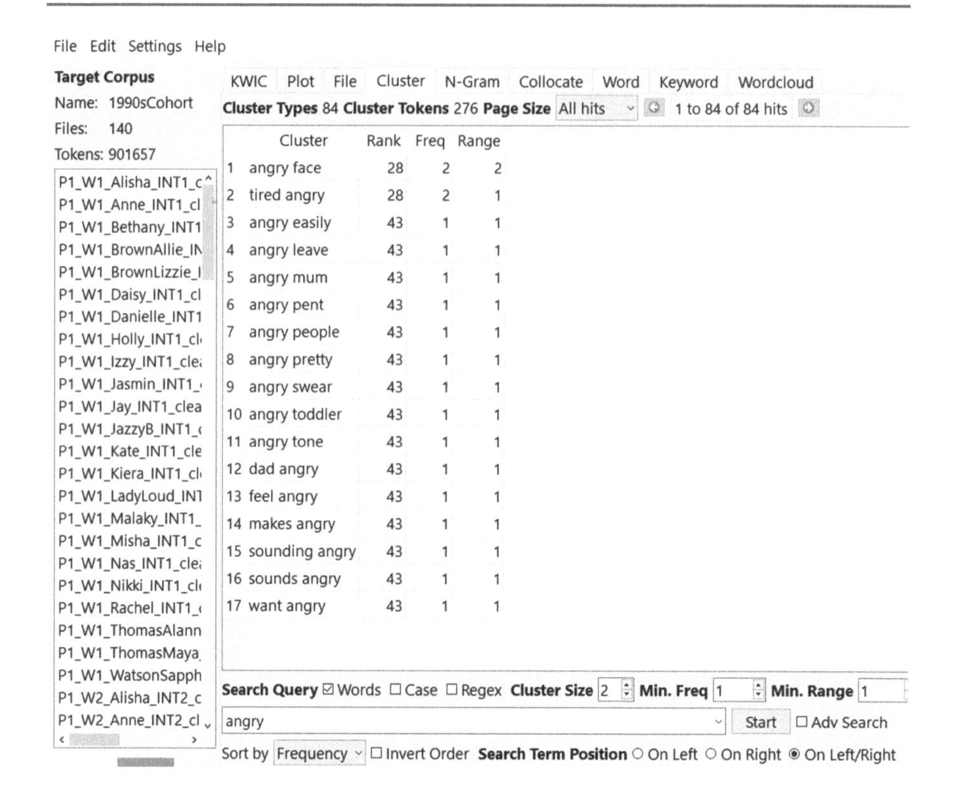

Fig. 5.6 Word clusters around 'angry'

Table 5.2 Document-term matrix illustrating 'bag of words' approach

	Angry	Let	Borrow	Shoes	Mess	Bit
Doc1	3	5	1	2	2	33
Doc2	0	7	1	0	1	15
Doc3	9	9	0	1	0	4

is intentional to convey what is often called the 'bag of words' approach. The analysis based on this matrix puts all the emphasis on the frequency of words and completely disregards the order in which they appear. It is not that the word order is lost but it is simply not used in the routines of computational analysis typical of topic modelling.

5.5.8 Latent Dirichlet Allocation (LDA)

What 'occurring together' means when starting from a document-term matrix is a high probability of being part of a particular topic and appearing in the same document as many other words associated with this topic. The mathematical assumptions and statistical procedures involved often start with 'Latent Dirichlet Allocation', to make best guesses about this. The procedures involved are working backwards from the assumption that a corpus of documents contains a set number of topics which have a particular distribution (e.g. topic 1 is 20% in total and topic 2 30%, etc.) but each topic may appear in greater or lesser proportions in any one document. Moreover, because each topic has its own vocabulary, each topic can be thought of as a distinctive basket of words. Each word has a particular probability of appearing in a particular topic basket.

The computer algorithms start with an arbitrary total number of topics which is set by the user. The process that then follows is described in equations in multiple web tutorials. Our account is a lay gloss which does not present or attempt to explain the mathematical formulae involved. The process starts with an assumed x number of topics across all documents and a random process of allocating words (or sentences or some other pre-defined unit of text) to topics. The random allocation then gets refined and further refined by multiple iterations of comparisons between documents that seek a resolution combining the proportion of words in a document belonging to a particular topic and the probability of each word appearing in a particular topic. Although the assumptions allow for words to appear in more than one topic, the computational procedures are seeking the best unique clustering of words discovered when all words are allocated according to their probability of clustering together in one of the x number of topics with a particular discovered distribution. The journal *Signs* has used LDA topic modelling to analyse content between 1975 and 2014 and provides a useful illustration of the technique (http://signsat40.signsjournal.org). In their published article and the associated supplementary material, Rao and Taboada (2021) provide a mathematically literate description of LDA topic modelling. Their worked example uses Python in an analysis of gender bias in how topics are presented in Canadian newspapers. Their code is publicly available on GitHub (https://github.com/sfu-discourse-lab/GenderGapTracker).

As is explained in more detail in Case Study 5.2, Nelson's analysis of focal concerns of different women's movement organisations combined identifying topics using LDA and comparing the outputs of different organisations using 'difference of proportions' analysis. Her discussion of LDA emphasises how the results are influenced by pre-processing decisions and the specified number of topics. Guided by the literature (Blei, 2012), her first attempts tried a range of numbers of topics from 20 to 200, finding that for her particular corpus 20 was too few as it combined several issues into one and anything over 100 was "not interpretable" (2014, p. 101). Her published work describes how she settled on 40 topics as the ideal number, as 50 resulted in "multiple topics on the same issue" (2020, p. 20). In addition to LDA, Nelson used one of a number of possible techniques for comparing the output of two different sections of the women's movement to further deepen her analysis of their difference. "Difference of proportions" analysis looks for "words with the largest positive and negative differences in a pair-wise comparison are the most distinctive words for each organization" (Nelson, 2020, pp. 15–16). She too takes the commendable step of making the computer code underpinning her published analysis available on GitHub (https://github.com/lknelson/computational-grounded-theory).

5.5.9 Visualisation

Computer packages for data analysis have routinely included data visualisation, forms of presenting data that facilitate 'seeing', interpreting, and communicating it. More ingenious forms of 'infographics' are regularly being added to the standard repertories of graphs, pie charts, and bar charts. Researchers working with text, using topic modelling and open-source programming such as R and Python, are often drawing on visualisation tools that others have already developed and/or are leaving a legacy of the code that they have used, as is the case in the work of Rao and Taboada (2021) capturing the differential prominence of men and women in particular topics covered in Canadian news outlets (Fig. 5.7).

However more basic techniques still have their place in helping to visualise text analysis. For example, turning word frequencies or 'keyness' into bar charts can aid comparative analysis. The bar chart in Fig. 5.8 draws on the analysis of gender and care for our big qual project and shows young men using words for parents, leisure activities, and school more frequently than young women who use more emotion words (close, love, annoy, hate, fight, upset). For results of the same techique with the older cohort see Davidson et al. (2019).

Word clouds, which translate the differential frequency of words in a text into different sizes of words, are a relatively simple visualising technique.

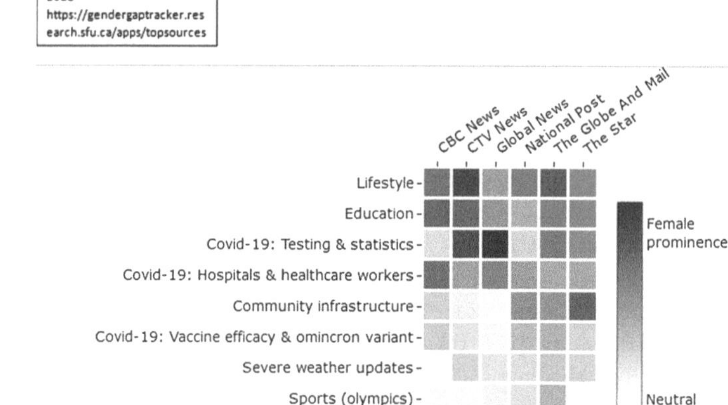

Fig. 5.7 Colour-coded visualisation

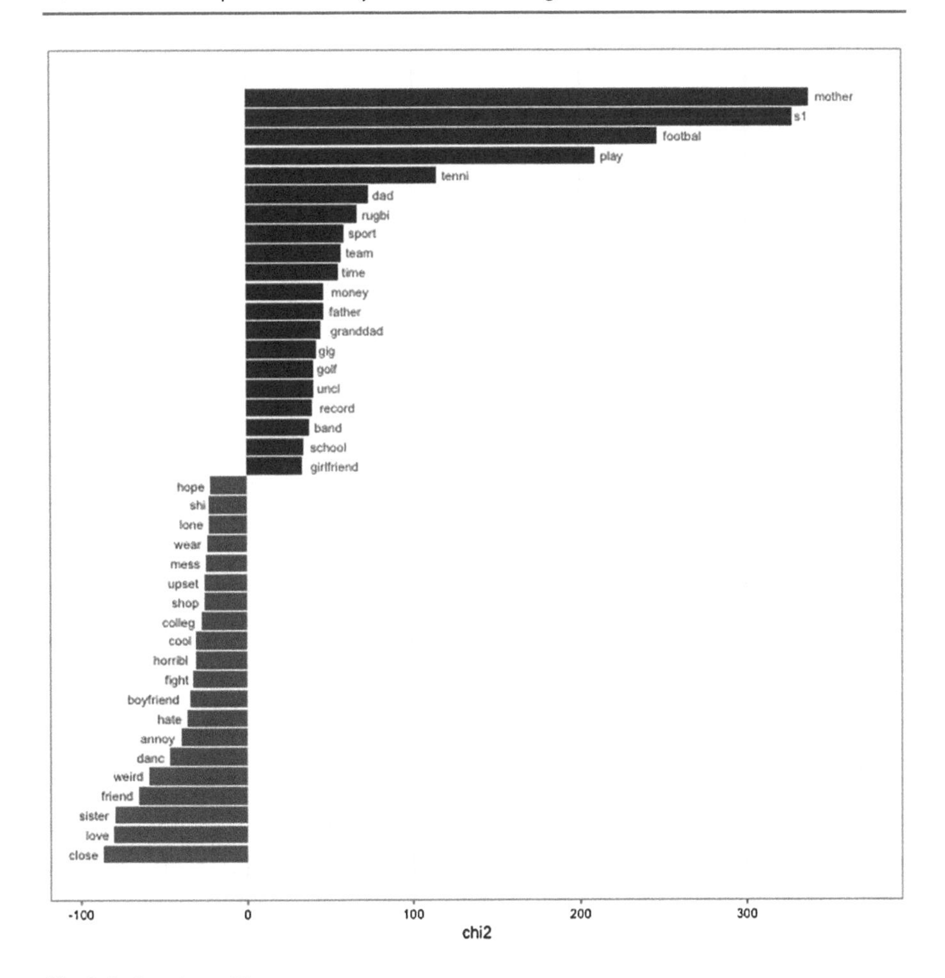

Fig. 5.8 Bar chart of keyness

As was noted earlier, commercial text analysis software, such as Leximancer, claiming to map topics without a steep learning curve often offer visualisations that facilitate 'seeing' and interpreting the data. Case Study 6.3 in the following chapter details our exploration of moving home across the life course using the breadth-and-depth method. Six data sets, housed in the Timescapes Data Archive, were pooled and reorganised into gender and generation groupings. Figure 5.9 shows the concept maps generated for 'all women' and 'all men' using Leximancer (v. 4.51). The software uses word frequencies and co-occurrence to create clusters of terms that are inclined to feature together in a text. Concepts are denoted by a dot, the size of which is determined by the concept's connectivity to the other concepts, and this is related to the frequency of occurrence. Concepts are grouped into themes. Themes deemed more salient are highlighted in hotter colours (reds, oranges) whilst those deemed less salient are represented by cooler colours (blues, green). These outputs cannot be compared to a simple word count/keyword frequency. Using two-sentence

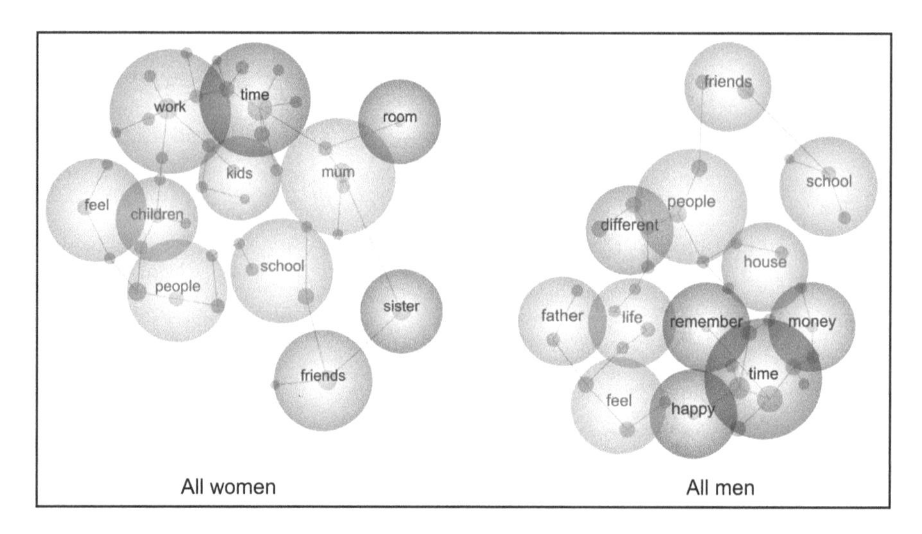

Fig. 5.9 Concept maps generated using Leximancer

segments, the concept count is likely to be greater than a simple keyword count as it is thesaurus-based and will include similar/related words. An evidence threshold must be met for a sentence block to be assigned to a concept. It is possible that a segment of text is assigned to a particular concept without containing the specific word. These maps, along with a range of other forms of output such as the ranking of concepts, can help guide the researcher where to dig deeper into the data.

As with conventional tables, being able to use and read a visualisation to better explain and display data requires an understanding of their construction.

5.6 Conclusion

The precise ways in which computational textual analysis might be used will vary with the questions the research hopes to address in bringing together a number of projects in a new data assemblage.

Management of known differences between projects predating the new data assemblage, such as in how data have been framed by the different questions asked or in the composition of the research participants, is likely to be an aspect of the analysis plan, for example shaping the use of comparison and perhaps modifying the necessary preparatory pre-processing of text for the chosen computational techniques.

However, it is hoped that the account of computer text analysis and text mining provided makes the systematic procedures and logics sufficiently transparent to provide a starting point for social science researchers who are unfamiliar with the techniques involved. The account also underlines the remoteness of the computational techniques from the conventions followed by human social scientists calling themselves qualitative researchers. This is in part an encouragement to embrace the

Table 5.3 Key terminology employed in text mining

Term	Meaning/definition
'Bag of words'	A starting point for analysis that reduces the entire 'corpus', data set of all text being analysed, to a list of the 'tokens' it contains and their frequency, with the assumption that frequency is indicative of meaning content and is more important than the order of words which may be disregarded
Keyword/ Keyness	'Keyword', a word that may represent a particular topic, is also given a more specific definition that makes reference to relative saliences or significance; 'keyness', words which occur with unusual frequency or infrequency in a target corpus in comparison to a reference corpus
LDA	Latent Dirichlet Allocation topic modelling is a set of computational procedures using frequency, proximity, and co-location of words and probability theory to indicate the best fit to X number of topics (with the number being specified by the user) across a corpus. Uses the assumption that a topic has its own distinctive basket of words, although some words may appear in more than one topic, and that X topics will have a particular distribution across the corpus but this may be different in each document
Lemmatisation	A process intended to reduce related forms of a word to a common base following a set of rules
Stemming	A process intended to reduce related forms of words to a common base using rules that mainly chop off certain word endings
Stop words	Words regarded as clutter, carrying no important meaning, and that are best removed from the text prior to analysis
Token	A small unit of text (usually a word) identified by a set of rules (that are usually intended to identify words)
Tokenization	The process of breaking a text down into tokens through the use of a programming language that enacts the specified rules
Topic	In text mining a 'topic' is essentially a cluster of words that frequently appear together and have a high probability of appearing together within a particular text or corpus
Vectorization	The representation of strings of text (e.g. sentences, documents) as numbers based on which tokens/words are present (1) and which tokens/words are absent (0) from the total set of tokens in the corpus

breadth-and-depth method as a multi-step iterative process rather than to stop at step two. It is also an encouragement to understand the affordances offered by computational textual analysis that open up ways of working with big qual, so that these can be embraced without false hopes of quick fixes and forewarning about the efforts involved in learning new skills and hands-on trial and error of developing competence. For a summary of terms that may be useful for researchers unfamiliar with text mining, see (Table 5.3).

5.7 Resources

Examples of text mining software

- Python
- https://www.nltk.org/
  https://scikit-learn.org/stable/
  https://pypi.org/project/pyLDAvis/

- R https://cran.r-project.org/web/packages/tm/index.html
  https://cran.r-project.org/web/packages/openNLP/index.html
  https://cran.r-project.org/web/packages/lda/index.html
- MALLET http://mallet.cs.umass.edu/topics.php.

Case Study 5.1 Extract from Clive Seale, Sue Ziebland and Jonathan Charteris-Black (2006). Gender, Cancer Experience and Internet Use: A Comparative Keyword Analysis of Interviews and Online Cancer Support Groups. ***Social Science & Medicine,*** 62(10), 2577–90

Comparative Keyword Analysis

Analysis employed Wordsmith software developed for corpus linguistics studies (Adolphs et al., 2004) to compare the relative 'keyness' of keywords, which are defined in the Wordsmith manual as 'words which occur unusually frequently in comparison with some kind of reference corpus' (Scott, 2005). This then is a purely quantitative conception of 'keyword', contrasting with qualitative conceptions such as that of Williams (1975). In such qualitative conceptions, a word is only deemed 'key' after a scholarly, interpretive investigation of its resonance within a system of ideas. A quantitative conception is by contrast purely mechanical, requiring the analyst to infer meaning after the identification of the word. In this respect, comparative keyword analysis is a more purely inductive approach than that of the qualitative analyst, who deploys inference at a much earlier stage. Interpretive analysis is still required, however, to identify meaningful clusters of keywords that describe important dimensions of difference.

In conventional corpus linguistics, the text of interest is compared with a large 'reference corpus' chosen to be broadly representative of general usage (for example, the British National Corpus). Stylistic, grammatical or other characteristics are elucidated by examining the words that occur more frequently in the text of interest than they do in the reference text. By contrast, the comparative key-word analysis reported here did not involve a general reference corpus. Instead, we identified keyword frequency in relevant texts compared with each other (e.g. breast cancer text compared with prostate cancer text), and then used this quantitative information to facilitate an interpretive, qualitative analysis focusing on the meanings of word clusters associated with keywords. Specifically, text produced by PWCs [people with cancer] with prostate cancer was compared with that produced by breast cancer PWCs, using the breast cancer text as a substitute 'reference corpus' and then the reverse was done. Extracts of some results of this comparison (for the internet forum text) are shown in the table below. The software in this instance identifies 'PSA' (Prostate Specific Antigen) and 'PROSTATE' as the two words that occur in the prostate cancer text most frequently, by comparison with the breast cancer text.

Thus 'PSA' occurs 1164 times in the prostate text (0.42% of all words in this text) but (unsurprisingly) only once in the breast text. An indicator of 'keyness' is given (the p-value of this, based on x^2 is also given by the software, but is not shown here since $p < 0.00000001$ for all keywords shown). In the latter half of the display 'negative' keywords are shown, indicating keywords in the breast cancer text.

Keywords in their contexts (KWIC lists) were examined, and Wordsmith was also used where necessary to examine clusters of words most frequently associated with keywords [...]. Keywords were thereby classified into meaningful categories (shared semantic fields) in a process analogous to the development of a coding scheme for the interpretive qualitative analysis of text. [...] For each comparative keyword analysis all keywords in the top 300 were examined. A word was excluded if it referred to the name of another person on the message forum, occurred less than 10 times, or where examinations of clusters showed it either to have meanings not significant for the analysis or highly variable (three or more) meanings (see examples at base of the table included in this case study). If examination of clusters indicated that a keyword should be associated with two of the coding categories it was included in both categories with a marker placed on it to indicate its status as a 'split' word. [...]

This enabled important and meaningful comparative aspects of these large bodies of text to be identified. This could be done in a more economical and potentially replicable manner than conventional qualitative thematic analysis based on coding and retrieval. At the same time, the more inductive approach to formulating coding categories allows for a greater openness to new findings than is allowed by some text analysis software programmes (e.g. Pennebaker et al., 2001) that rely on pre-formulated dictionaries to allocate words to pre-specified categories. An additional conventional thematic content analysis of relevant interview text was done to identify gender differences in internet use reported in interviews.

Meaningful words in top 25 positive and 25 negative keywords: Comparison of people with prostate and breast cancer in web forums

Keyword	Freq.	%	RC.Freq.	RC. %	Keyness
Positive					
PSA[a]	1164	0.42	1		4141.74
Prostate	1080	0.39	28		3606.20
RP[a]	339	0.12	2		1186.17
PC[a]	377	0.14	30		1142.48
Gleason	285	0.10	0		1017.19
PCA[a]	299	0.11	6		1010.35
Regards	393	0.14	111		912.12
RT[a]	290	0.10	44		790.90
Brachytherapy	151	0.05	0		538.87
Men	286	0.10	180	0.01	465.20
Score	154	0.06	31		393.73
Catheter	128	0.05	14		370.49
Urologist	89	0.03	0		317.60
Bladder	110	0.04	15		306.32
Prostatectomy	83	0.03	0		296.18
Hormone	214	0.08	193	0.01	271.52
Dad	177	0.06	125		267.95
Negative					
Care	82	0.03	1252	0.09	−136.78
Nodes	40	0.01	905	0.07	−144.31
Arm	8		586	0.04	−159.25
Feel	332	0.12	3260	0.24	−170.20
My	2852	1.03	18,370	1.34	−183.42
Mets	8		673	0.05	−189.06
They	824	0.30	6577	0.48	−189.11
Think	386	0.14	3800	0.28	−199.65
Mum	7		706	0.05	−205.98
X[a]	34	0.01	1085	0.08	−215.86
Me	1127	0.41	8757	0.64	−228.16
Women	22		1060	0.08	−253.44
I'm	303	0.11	3508	0.26	−256.03
Hair	6		834	0.06	−256.91
Lump	9		1008	0.07	−299.87
Love	55	0.02	1681	0.12	−326.67
Her	113	0.04	2797	0.20	−476.60
She	167	0.06	3330	0.24	−479.34
I	8174	2.95	53,671	3.91	−624.75
Breast	40	0.01	3653	0.27	−1045.87
Chemo	16		3578	0.26	−1169.09

Excluded words from positive list: 'the' and 7 names of other forum participants.
Excluded words from negative list: 'it', 'really', 'don't', 1 name of another forum participant.

[a]Concordance and inspection of clusters showed majority of X's in breast cancer text to be kisses. PSA = Prostate specific antigen, RP = Radical prostatectomy, PC and PCA = Prostate cancer, RT = Radiotherapy.

(pages 2581–2582 and Table 2).

▶ Case Study 5.2 From the History of Feminist Ideas to 'Computational Grounded Theory'

Summary and extracts from Nelson (2021). Cycles of Conflict, a Century of Continuity: The Impact of Persistent Place-Based Political Logics on Social Movement Strategy. *American Journal of Sociology* 127(1):1–59, 202 and Nelson, L. (2020). "Computational Grounded Theory: A Methodological Framework." *Sociological Methods & Research,* 49(1), 3–42.

The focus of Laura Nelson's historical work is feminist analysis, claims and debates within the USA women's movement in the two different periods that are often labelled 'first' and 'second' wave feminism (2014, 2020). She examines the 'cognitive frameworks' of the movement in Chicago and New York having identified the relevant organisations and network of actors in each place in each period. She painstakingly assembled data from multiple archives that enabled both regional comparison between the cities and temporal comparison within each city between 'first' and 'second' waves. From these data she constructs an overview historical narrative drawing on multiple historical sources. This includes a qualitative network analysis tracing links between people and organisations within each city, detailed reading of the views expressed in published writings of their leading figures and computational textual analysis of the literature produced by the key organisations. The preparatory work often included having to convert scanned copies of archive documents into text. In combination the elements of her analysis demonstrated that in both periods, a different set of agendas, priorities and political styles persisted between Chicago and New York. This enabled Nelson to conclude that the ebbs and flows within the US movement are not appropriately categorised as 'first' and 'second' waves since the blend of different viewpoints is long standing, persistent and explicable in terms of the social context in which key actors are embedded.

The corpus on which Laura Nelson conducts computational analysis consisted of the key publications of the four feminist organisation that she identified as the leading organisations in Chicago, Hull House and Chicago Women's Liberation Union [CWLU], and in New York, Heterodoxy and Redstockings, in each of the relevant periods, summarised below (Nelson, 2021, p. 14).

Summary of Corpus

City/Organization	Wave	Publication	Years	Word Count	Page Count	Article Count
Chicago:						
Hull House	First	*Bulletin/Year Book*	1900–1917	200,747	357	~56*
CWLU	Second	*Womankind*	1971–73	303,306	364	. . .[†]
New York City:						
Heterodoxy	First	*Masses*	1911–17	70,393	78	67[‡]
Redstockings . . .	Second	*Notes from the First Year, Notes from the Second Year,* and *Feminist Revolution*	1968–75	264,248	332	76[‡]

* Hull House did not publish a table of contents or an index, so I calculated the approximate number of articles by multiplying the number of articles in one issue by the number of issues included in the analysis.

[†] *Womankind* did not produce a table of contents or an index, and it was difficult to distinguish what counted as a distinct article.

[‡] See app. C for a list of articles.

She used two forms of computational text analysis. Probabilistic topic modelling, that is Latent Dirichlet Allocation [LDA] modelling and related forms of Structural Topic Model were applied to the whole corpus to identify 'topics' within the corpus. She also used difference of proportions to compare the pattern of key terms between pairs of organisations. These two steps might be worked through using topic modelling first and then difference of proportions or using difference of proportions first and then topic modelling. Each has slightly different logics. In the following extract she describes the latter: "In the first step of my analysis, the pattern detection step, I used a combination of difference of proportions (a lexical selection technique) and STM (a structural topic modeling algorithm) on the literature produced by four core women's movement organizations [...] I began the analysis with four pairwise difference of proportion calculations to extract the most defining words for each pair of organizations. [...] These words can be analyzed to suggest patterns within the corpus (which I do below). To further explore the themes addressed in this literature and to categorize the text into those different themes, I followed this analysis with structural topic modeling. To determine which model was best for my corpus, I ran four topic models, with 20, 30, 40, and 50 topics, respectively. Examining the top weighted words for each model, I found some of the topics in the 20-topic model combined two issues into one. For example, one of the topics in the 20-topic model had the top weighted words: car, can, women, doctor, gonorrhoea, and infect. I found these words combined the issue of car maintenance and the issue of sexually transmitted infections, meaning the 20-topic model had too few topics. Conversely, I found the 50-topic model produced multiple topics on the same issue. For example, the top three weighted words in one topic in the 50-topic model were class, year, and art, while in another topic they were hullhouse, children, and school, and in yet another they were school, class, and boy. I interpreted these three topics to all be about one issue: different types of classes offered by Hull House. For my purposes, I was looking for more general topics than the specific types of classes Hull House taught, so I determined the 50-topic model was too specific. The 40- topic structural topic model, alternatively, produced topics that were comfortably distinct from one another, yet general enough to be interpretable for my purposes, so I used this model for my analysis" (2020, p. 20). An illustrative table shows the 3 top topics for each organisation, indicating the percentage of the organisation's literature that is aligned with the topic and the 20 top weighted or most distinctive words that the model has allocated to that topic. For example, Nelson identifies the top Hull House topic which accounts for 28% of their literature as 'public institutions' which includes words referring to school, housing and work. As Nelson explains, it is up to the researcher to interpret the words and name the topic, a process that also involved reading representative documents for each topic.

To achieve a 'computational grounded theory', Nelson (2020) advocates using the computer identification, comparison and weighting of topics to select documents for the careful detailed reading and human coding of qualitative data analysis. Then, having reached further conclusions on the basis of a qualitative analysis to return to computational techniques in order to conduct further checks on generalisability across the data. This final step requires the researcher to identify one or more additional independent computational techniques to those already used, that will test at least an aspect of the conclusions reached. In the case of her own historical research on the US Women's movement, she drew on computational techniques for analysis of word hierarchies drawing on WordNet (Princeton University 2010) and a crowd-sourced dictionary to measure the level of specificity and abstraction of words in texts originating from the women's movement in Chicago versus texts originating in New York. In addition, she used a program designed for 'name entity recognition' that helps to distinguish between named individuals, organisations, locations and other entities in order to test whether Chicago feminist generated texts more frequently name organisations and New York feminist generated texts more frequently name individuals. She notes 'There is a certain amount of creativity in this last step, however, and successfully applying this step to other research projects requires general knowledge of the range of text analysis and natural language processing tools available.' (2020, 33).

References

Adolphs, S., Brown, B., Carter, R., Crawford, P., & Sahota, O. (2004). Applying corpus linguistics in a health care context. *Journal of Applied Linguistics, 1*(1), 9–28.

Anthony, L. (2009). Issues in the design and development of software tools for corpus studies: The case for collaboration. In P. Baker (Ed.), *Contemporary corpus linguistics* (pp. 87–104). Continuum.

Anthony, L. (2022). *AntConc (Version 4.1.2) [Computer Software].* Waseda University. http://www.antlab.sci.waseda.ac.jp/

Blei, D. M. (2012). Probabilistic topic models. *Communications of the ACM, 55*(4), 77–84.

Bonzanini, M. (2016). *Mastering social media mining with python.* Packt Publishing Ltd..

Bryman, A. (2016). Social research methods (5th ed.). Oxford University Press.

Davidson, E., Chun-ting Ho, J., & Jamieson, L. (2019). *Computational text analysis using r in big qual data: Lessons from a feasibility study looking at care and intimacy.* National Centre for Research Methods. Big Qual Analysis Resource Hub. Retrieved 2022.

DiMaggio, P. (2015). Adapting computational text analysis to social science (and vice versa). *Big Data & Society, 2*(2), 2053951715602908.

Graham, S., Weingart, S., & Milligan, I. (2012). *Getting started with topic modelling and MALLET: UWSpace.*

Grimmer, J., & Gary, K. (2011). General purpose computer-assisted clustering and conceptualization. *Proceedings of the National Academy of Sciences, 108*(7), 2643–2650. https://doi.org/10.1073/pnas.1018067108

Jockers, M. L. (2014). *Text analysis with R for students of literature.* Springer.

Krippendorff, K. (2019). *Content analysis: An introduction to its methodology.* Sage.

Lutz, M. (2001). *Programming python.* O'Reilly Media, Inc.

Mills, K. (2019). Big data for qualitative research (1st ed.). Routledge.

Nelson, L. K. (2014). The power of place: Structure, culture, and continuities in US women's movements. UC Berkeley.

Nelson, L. K. (2020). Computational grounded theory: A methodological framework. *Sociological Methods & Research, 49*(1), 3–42.

Nelson, L. K. (2021). Cycles of conflict, a century of continuity: The impact of persistent place-based political logics on social movement strategy. *American Journal of Sociology, 127*(1), 1–59.

Pennebaker, J. W., Francis, M. E., & Booth, J. R. (2001). *Linguistic inquiry and word count (LIWC): A computerized text analysis program.* Erlbaum Publishers. (See also Pennebaker, J. W., Booth, R. J., Boyd, R. L., & Francis, M. E. (2015). *Linguistic inquiry and word count: LIWC2015.* Pennebaker Conglomerates (www.LIWC.net)).

Porter 2/Snowball Stemmer (includes suggested stop word list). (n.d.). http://snowball.tartarus.org/algorithms/english/stemmer.html

Porter Stemmer. (n.d.). https://tartarus.org/martin/PorterStemmer/

Python Natural Language Toolkit. (n.d.). https://www.nltk.org/

Rao, P., & Taboada, M. (2021). Gender bias in the news: A scalable topic modelling and visualization framework. *Frontiers in artificial intelligence, 4*, 664737–664737.

Scott, M. (1997). PC analysis of key words—And key key words. *System, 25*(2), 233–245.

Scott, M. (2005). *Wordsmith tools 4.0.* Oxford University Press.

Seale, C., & Charteris-Black, J. (2008). The interaction of age and gender in illness narratives. *Ageing and Society, 28*(7), 1025–1045.

Seale, C., & Charteris-Black, J. (2010). Keyword analysis: A new tool for qualitative research. In I. Bourgeault, R. Dingwall, & R. De Vries (Eds.), *The SAGE handbook of qualitative methods in health research* (pp. 536–665). Sage.

UCREL Semantic Analysis System/ http://ucrel.lancs.ac.uk/usas and the English tagger is http://ucrel-api.lancaster.ac.uk/usas/tagger.html

Wiedemann, G. (2016). *Text mining for qualitative data analysis in the social sciences* (Vol. 1). Springer.

Williams, R. (1975). *Keywords: A vocabulary of culture and society.* Fontana.

'Test Pit Sampling': Preliminary Analysis 6

6.1 Introduction

The previous two chapters outlined the processes of conducting an enquiry-led overview of relevant qualitative research (step one) and computer aided scrutiny across the breadth of selected data collections to assess what merits closer investigation (step two). This chapter focuses on a critical juncture in the breadth-and-depth method: the shift from gaining a sense of the breadth of a data assemblage to starting to look at the material in more depth. In essence, step three of the method is concerned with the analysis of multiple small samples of data that, on the basis of the recursive thematic mapping conducted in step two, show promise of containing data pertinent to your research topic or question(s). Drawing on the archaeological metaphor originally detailed in Chap. 2, this step can be likened to digging shallow 'test pits' in areas of the data landscape which, based on the outcomes of the 'geophysical surveying' or text mining, appear of salience to your research aims, objectives, and questions. The aim of this chapter is to guide you through the procedures and decision-making involved in this step. In so doing, we will:

- detail the process of moving from breadth to depth
- guide you through the steps of digging deep enough into the data to see whether there is material of interest
- discuss key issues such as sample size and sampling logic
- explain how to undertake cursory readings of short extracts to gain a clear sense of whether the material is of relevance
- provide illustrative examples from a range of research projects detailing different approaches to the process
- offer some suggestions for key issues to consider
- provide links to useful online resources

© The Author(s), under exclusive license to Springer Nature Switzerland AG 2023

S. Weller et al., *Big Qual*, https://doi.org/10.1007/978-3-031-36324-5_6

6.2 Moving from Breadth to Depth

Step three of the breadth-and-depth method focuses specifically on the process of moving from gaining a sense of the breadth of the material in your assemblage to digging a little deeper, sampling the data, and undertaking the preparatory work necessary to enable more in-depth interpretive analysis in step four. At this point, the data you are working with may take a number of forms. When working with software that enables visualisation, you are likely to have outputs in the form of two-dimensional concept maps founded on relationships, frequencies, and correlations (Lewins & Silver, 2020). Figure 6.1, for instance, is the output from the text-mining software Leximancer (v4.51). Concepts are denoted by a dot, the size of which denotes it's relative connectivity to the other concepts. A collection of related concepts constitute a theme. The most dominant concept determines the theme's label (Cretchley et al., 2010). The strength of the themes is colour-coded with hotter colours (i.e. red, orange) relating to the strongest themes and cooler colours (i.e. blue, green) denoting the least salient (Gapp et al., 2013). The resolution of the map can be altered to reveal more or less salient themes, and as you do so concepts may be re-allocated.

When using more basic procedures than topic modelling or freeware, forms of visualisation may be more restricted to bar charts or tables, as shown in the examples and case studies in Chap. 5: Tables 5.2 and 5.3 of keyness using Antconc, the table comparing meaningful words used by those living with prostrate and breast cancer in Case Study 5.1 by Seale and co-authors, and the tables that can be found in the work of Nelson summarised in Case Study 5.2. Nelson uses R and Python in topic modelling to identify 'topics' within her corpus for further qualitative investigation, but visualisation of her output is in the form of a very large table showing the top topics for each organisation, indicating the percentage of the organisation's

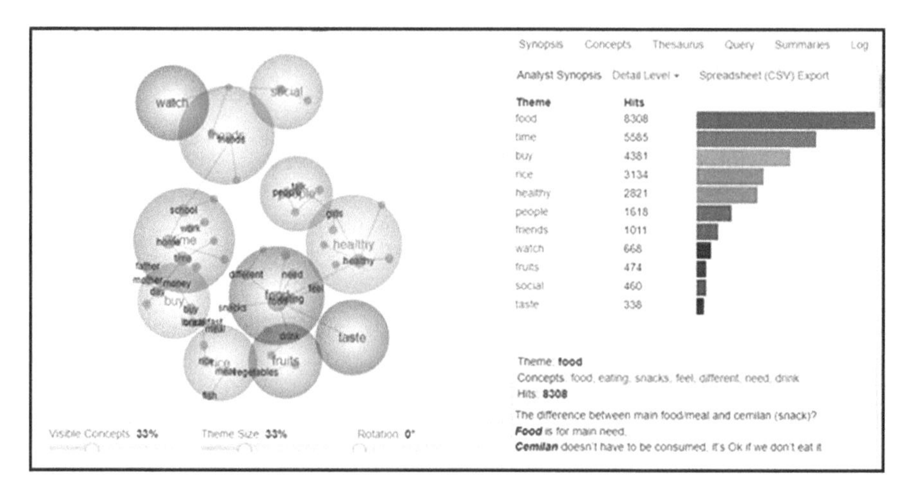

Fig. 6.1 Example output from the text-mining software Leximancer (for details of the analysis in which this was used see Neufeld et al., 2022)

literature that is aligned with the topic and the 20 top weighted or most distinctive words that the model has allocated to that topic (Nelson, 2021, Table 4).

These outputs offer an overview of what may be the key concepts and themes identified in the data. However, as the examples discussed in Chap. 5 illustrate, preliminary outputs often require considerable further investigation. The examples discussed in the previous chapter also prefigure the 'test pit' sampling discussed in this one. For example, further investigation showed that not all the differences in keyness shown in Table 5.2 represented meaningful differences between young men and women. This further investigation required reading samples as 'test pits'. Within Antconc this was done by switching between 'keyness' and KWIC, keyword in context, in order to select and read through a sample of extracts with the keyness word. Given our interest in potential convergence in vocabularies and practices of intimacy between men and women over time, we were particularly alert to differential usage of relational and emotional terms. As discussed in Chap. 5, preliminary examination of 'keyness', for example, revealed the relatively different frequency of use of some relational terms such as 'mum' and 'friend' and the word 'play' by boys and 'close' and 'talk' by girls. The latter suggested possible continuity of gender differences in the intimacy of friendship relationships for further in-depth investigation. We examined extracts of text in more detail using each of these words with the KWIC tool, to show the keyword in context, and then to select samples in which we read the sentences around the keyword. This revealed that the differences in use of 'mum' and 'friend' were a false start being, at least in part, an artefact of our own anonymisation process. Figure 6.2 shows the output of the KWIC tool displaying uses of 'mum' ordered by the two words to the left and the word to the right. Scanning such an output gives an initial idea of how the word is used and the full

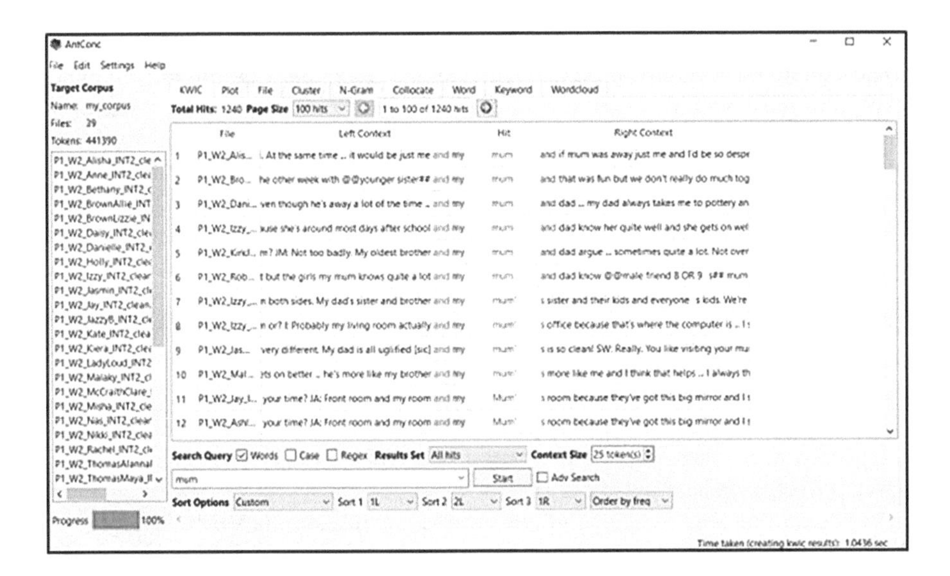

Fig. 6.2 Antconc example: KWIC as a starting point for test pit sampling

text is just one click away in each case. This makes it possible to select a sample for more careful scrutiny and to decide whether it was worthy of further exploration relative to our research questions. Our experience involved ruling out multiple terms during this step.

As Nelson explained in Case Study 5.2, it took various iterations with different numbers of topics before arriving at the analysis of the different concerns of particular feminist organisations. Her judgements about the correct number of topics and how to interpret the topics, name them, and decide whether they were indeed representing a difference in content meaning again involved reading samples.

Having done test pit sampling, you may ask whether this level of analysis will suffice. In designing the breadth-and-depth method, however, we felt that, to work with qualitative integrity further, more-in-depth analysis and interpretation was essential (hence step four). Importantly, we do not regard computational analysis as the endpoint. Instead, by combining the insights offered by text-mining techniques with the benefits of in-depth analytic work, we propose a form of large-scale qualitative analysis which unites quantitative and qualitative approaches in active dialogue resulting in both extensive coverage and intensive illumination (Davidson et al., 2019, p. 368). In so doing, and as discussed in Chap. 2, we sought to develop an approach to big qual that respects the principles underlying conventional qualitative data analysis, ensuring that "the distinctive order of knowledge about social processes that is the hallmark of rigorous qualitative research, with its integrity of attention to nuanced context and detail" is retained (Davidson et al., 2019, p. 365; see also Bruns, 2013).

6.3 The Metaphorical Foundations of Test Pit Sampling

As discussed in Chap. 2, we found the use of metaphor, and in particular archaeological field investigations, helpful in thinking about how we might combine exploring the breadth of a vast data landscape with examining the material in depth. The breadth-and-depth method commences with a process that resembles aerial reconnaissance whereby the researcher flies systematically over a data landscape to gain a broad sense of the nature of the material available. This is followed by geophysical surveying across the composite data set or assemblage, working only at the surface level to pinpoint features, for instance data of relevance to a research question or project. It is in step three, where the focus turns to conducting preliminary analysis, that we start to probe a little deeper, breaking through the surface layer of the ground by digging shovel test pits or small excavation holes to assess whether these features are of any value (see Fig. 6.3) (Richards, 2008). Where the archaeologist's trowel is placed is dependent on the outcomes of the geophysical surveying. Digging a test pit involves working by hand. Using a small shovel, each layer of soil is examined methodically with the aim of uncovering artefacts that help indicate the site's use-value. In a similar vein to the systematic arrangement of test pits across a site of archaeological excavation, step three requires the use of a sampling strategy. Furthermore, field archaeologists sift

Fig. 6.3 Archaeological metaphor and step three of the breadth-and-depth method. (Illustration created by Chris Shipton https://www.chrisshipton.co.uk/)

through the soil in test pits in a similar manner to a researcher sifting through small extracts of text. Features of interest are documented and perhaps categorised. Importantly, the aim is not to carry out in-depth interpretative work at this stage but simply to determine whether the test pit contains material of interest in preparation for deeper excavations in step four. If a test pit does not yield data of relevance, then the researcher returns to the geophysical surveying (step two) to locate alternative sites within the data landscape to explore. You may also need to revisit step one to seek additional or alternative data sources.

6.4 The Logic of Identifying Samples and Choosing Cases

The data extracts identified in step two will need to be selected for usefulness and salience to determine areas worthy of further investigation. Decisions about the quantity or volume of extracts to explore evoke similar responses to those concerning the size and nature of qualitative samples more generally, the common response being 'it depends'. It depends on epistemological and methodological questions about the nature and purpose of the research, on practicalities of time and resources, and also on the conventions and judgement of the epistemic community in which the researcher(s) is (are) located (Baker & Edwards, 2012). These sorts of key considerations are just as applicable to Qualitative Secondary Analysis (QSA) and big qual as they are to primary qualitative generation of research participants. For step three of the breadth-and-depth method, the strategy focuses on the approach taken to determining how many and which extracts are taken forward for a greater depth of analysis in step four.

Discussions of strategies informing the selection or choice of units for inclusion in research are usually considered in relation to the identification and recruitment of participants in primary research, covering many different approaches. These various strategies can be read about in any good qualitative methods text. Here we are concerned with logics that may also be applied in secondary research, to explore and identify material in preparation for embarking on step four. The various approaches may be understood as sampling that is driven by generating conceptualisation or by pragmatism, or as choosing cases within a realist framework.

If you are selecting extracts from interview transcripts within what is often referred to as a theoretical sampling logic, then you are looking to discover or construct concepts and knowledge from the data in an empirical, grounded fashion. The identification of extracts to sample in this logic is gradualist. It might involve the breadth-and-depth researcher starting with a few extracts, using them to generate some emergent ideas about what is of interest, returning to the body of extracts to sample comparatively, looking for whether the ideas hold up in similar and in distinct contexts (e.g. across different generations), and continuing in that fashion until a general theory emerges. In contrast, a pragmatic approach to sampling for extracts starts from purposeful judgements about what are information-rich examples that will provide insight into the key issues that the researcher has identified as important. You are using your existing knowledge about the topic and the implications of what may be key aspects of, say, participants' characteristics and experiences (such as gender and generation) to direct your sampling. The driver for identifying extracts for further investigation here would be that they are cases that provide explanations as to what is causing a situation or interaction, in what contexts, and with what outcomes. In this framework the breadth-and-depth researcher is strategically and purposefully choosing cases on the basis of their causal theories and contextual concepts, within the constraints of the data that they are working with.

The three strategies briefly reviewed here are simplified characterisations in what is a complex field, with many overlapping grey areas between approaches as well as within different branches of a single category. But what the approaches all have in common is that they are underpinned by an epistemology, by a particular understanding of the logic of the relationship between evidence and ideas in how we may generate knowledge about the social world. The breadth-and-depth researcher needs to be clear about which logic they adopt and why in identifying or choosing the transcript extracts they work with.

6.5 The Transition from Step Three to Step Four

Outputs generated from step two shape where you might decide to dig into the data landscape. You may wish to focus on the most salient keywords and concepts of interest identified through your recursive thematic mapping. For instance, as detailed in Case Study 6.1, in a similar fashion to the breadth-and-depth method, Zaitseva and Finn (2021) combined text-mining techniques to gain a sense of the breadth of data generated via open responses to the National Student Survey (NSS) in their higher education institution (HEI), with thematic analysis to explore depth. They employed the text-mining software Leximancer for the breadth analysis and, in deciding where to dig deeper, focused on concepts with a high relevance score and high number of instances as they were keen to identify topics and issues of importance to students. Using sentiment analysis, they also placed emphasis on concepts that had a higher probability of being mentioned unfavourably, allowing the team to gain a sense of aspects of student experience that respondents felt required improvement. The transition from breadth to depth was therefore guided by the focus of the research on understanding issues of salience to students.

Case Study 6.2 outlines a more a pragmatic approach where issues of importance had, to some extent, been identified through analysis of other data sources (Neufeld et al., 2022). The team employed the breadth-and-depth method to analyse 11 data sets from projects focusing on food choice in transition and young people's autonomy and agency. The pooled data sets were reorganised into new groupings delineated by age (young people and adults) and the type of local food environment in which participants resided. Concept maps were generated for each of these groupings and, along with details about the ranking of themes, were used to help identify issues of salience, with a particular emphasis on agency and autonomy. Connections between concepts were explored by running a series of Boolean queries (an outline of Boolean operators is available in Table 3.2). Some of these relationships were foreseeable given the nature of the original projects. The team were, however, keen to explore connections between less salient themes, and between those that featured in the concept maps of some groupings and not others, for example the presence/absence of discussions about the availability and consumption of 'snacks' or 'street food' in different food environments. The 'test pits' were, therefore, the outcome of the Boolean queries.

6.6 Conducting Cursory Readings

The next task in step three is to conduct cursory readings of short data excerpts to gauge whether the material is of relevance. At this stage you will be exploring, relatively quickly given the volume of data, samples of 'themes' using extracts of text data containing the keywords or concepts identified in step two. For many qualitative researchers, undertaking brief readings and not engaging in detail with the minutiae of the narrative is likely to prove challenging, especially given that many approaches to qualitative analysis encourage immersion in the data. It is important

anyway	SW	look
called	Yeah	moment
cause	able	old
course	coming	probably
cos	doing	round
erm	down	saying
fact	everything	stuff
inaudible	fact	take
laughs	goes	things
name	whole	thought
obviously	half	try
suppose	having	used
S2	long	

Fig. 6.4 Examples of keywords deemed to have ambiguous meaning. (From the 'Moving home across the life course' analysis outlined in Case Study 6.3)

not to engage in deeper readings at this stage since multiple features of interest should be investigated to justify where to undertake deeper analysis. Moreover, given the probable volume of the material you will be handling, more in-depth readings are likely to be impractical. Case Study 6.1, by Zaitseva outlines a large-scale analysis of the open questions in the NSS completed by students at their HEI (Zaitseva & Finn, 2021). The team worked with text totalling approximately 80,000 to 100,000 words. In Case Study 6.2, a pooled breadth-and-depth analysis of 11 data sets, all focused on young people and food choices in different international contexts, generated 500 pages of excerpts to study. Finally, in Case Study 6.3 a focus on moving home across the life course using six qualitative longitudinal research (QLR) data sets from the Timescapes Data Archive, resulted in 136 pages of material to examine. In all three case studies, this work was completed by small teams in a relatively short space of time.

In essence, the goal of this step is to understand just enough of the extract and the context in which it was discussed to decide whether it is worthy of inclusion in step four. In some instances, the decision will be relatively straightforward in that whilst the outputs from step two allude to a relationship between a set of keywords or concepts, this may be happenstance with the words actually corresponding to responses to different questions that feature in close proximity in a transcript. Alternatively, the keywords may have ambiguous or multiple meanings that are not of relevance to your research (Fig. 6.4).

In some instances, it is likely to be necessary to read more of the text surrounding an extract to glean a clearer sense of the meaning prior to determining its inclusion or exclusion. Figure 6.5 shows how, on the Leximancer platform, a data extract (highlighted in green) can be explored within the wider context in which it is situated within a text. Furthermore, even with preliminary analysis, it is important to keep the wider context in which the data was generated in mind, including knowledge of the project meta narratives and case metadata. Figure 6.5 also shows that

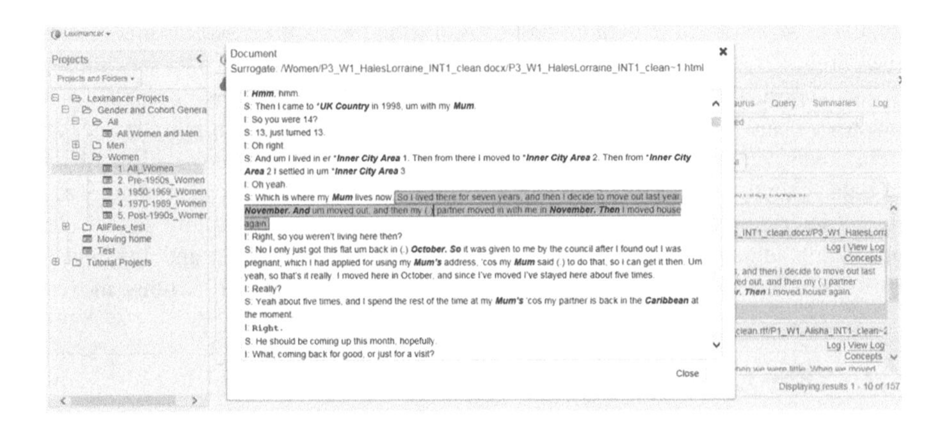

Fig. 6.5 Viewing extracts-in-context (see also Case Study 6.3)

with clear file-labelling (see Chap. 3) it is possible to view from which participant, and which data set the extract originates.

The length of the excerpts with which you will work is determined largely by the nature of the data. In our experience, extracts from interviews with individuals are usually within the region of 150 to 200 words, whilst those from focus groups tend to be shorter. The amount of text needs to be enough to incorporate several full sentences and to garner a clear sense of the context in which the keyword or concept is embedded. In practice, and depending on the nature of the data, you may find that shorter extracts suffice. For instance, the data assemblage described in Case Study 6.2 included transcripts from interviews and focus group discussions. In the case of the latter the interlocutors and topics often altered in quick succession and it was, therefore, only necessary to include shorter extracts. Any keywords or concepts that seem ambiguous should be eliminated. For consistency, it is advisable to log meticulously your rationale for including and excluding data, especially as this is a time-consuming process unlikely to be completed in one go.

If you already have clearly defined research questions, some keywords or themes may be substantively or theoretically more eye-catching, and these are likely to prove to be a useful starting point. Alternatively, you may be taking a more exploratory approach investigating further features that appear dominant in the data landscape. You may be more interested in what appear to be anomalies or more unusual occurrences. Prior to moving on to step four, it can also be fruitful to sample keywords or concepts that seem less striking but are of potential value to your analysis. Equally, and as demonstrated in Chap. 3, if a test pit does not result in any material of interest, then you will need to return to step two to conduct further mapping. Step three should, therefore, be regarded as a recursive process involving multiple iterations as the cursory readings are likely to reveal irrelevant material.

Based on your sampling strategy and any other features of your research design, we also recommend that you start to organise any extracts that you feel might prove valuable for more interpretative work in step four. For instance, Case Study 6.3

details how we used the breadth-and-depth method to explore participants' narratives around moving home across the life course. We drew on six data sets housed in the Timescapes Data Archive. The data assemblage comprised in-depth QLR interviews, and in line with the focus on exploring change and continuity over time, the extracts were organised by unit or case. A unit or case comprised multiple waves of data generated with an individual or a small group (i.e. couples, siblings, families) if interviewed together. We then organised the units or cases into gender and generation cohort groupings in order to start to explore similarities and differences between/across cohort generations with a focus on those 'left behind' in home moves.

6.7 Key Areas for Consideration

The following questions may help guide your thinking through step three.

Prior to undertaking step three

- What is the nature of your outputs from the recursive surface thematic mapping of step two?
- How will you decide which Boolean queries to run? How will the output from step two inform this process?
- How will you determine your sample?
- What sampling logic will you employ?

During the process of conducting preliminary analysis

- How will you manage a large volume of extracts? What software might you use?
- What will you keep in mind whilst conducting your cursory readings? Will you be completing these individually or collectively with colleagues?
- How will you decide which extracts to retain and which to dismiss?
- Do you need to return to step two to undertake further recursive surface thematic mapping?

Preparing for step four

- How might you organise your selected extracts?
- After selecting extracts to retain, how will you prepare these for more in-depth analysis?
- What approach(es) to in-depth analysis might be appropriate for your research endeavour?

6.8 Resources

The following papers provide further details about approaches to sampling and the place of theory in the breadth-and-depth method.

Baker, S. E., & Edwards, R. (2012). *How many qualitative interviews is enough?* Discussion Paper. NCRM. https://eprints.ncrm.ac.uk/id/eprint/2273/

Edwards, R., Davidson, E., Jamieson, L., & Weller, S. (2021). Theory and the breadth-and-depth method of analysing large amounts of qualitative data: A research note.*Quality & Quantity, 55*, 1275–1280.

Edwards, R., Weller, S., Jamieson, L., & Davidson, E., (2020). Search strategies: Analytic searching across multiple data sets and with combined sources, In K. Hughes, & A. Tarrant (Eds.), *Qualitative secondary analysis* (pp. 79–100). London.

The following blog posts, available from the Big Qual Analysis Resource Hub (https://bigqlr.ncrm.ac.uk/), provide examples of techniques for moving from exploring the breadth of a large qualitative data set to the depth.

Davidson, E., Chun-ting Ho, J., & Jamieson, L. (2019). Computational text analysis using R in Big Qual data: lessons from a feasibility study looking at care and intimacy. *Big Qual Analysis Resource Hub.* http://bigqlr.ncrm.ac.uk/2019/03/21/post27-dr-emma-davidson-justin-chun-ting-ho-and-prof-lynn-jamieson-computational-text-analysis-using-r-in-big-qual-data-lessons-from-a-feasibility-study-looking-at-care-and-intimacy/

Fadyl, J. (2018). Seeing the changes that matter: QLR focused on recovery and adaptation. *Big Qual Analysis Resource Hub.* http://bigqlr.ncrm.ac.uk/2018/10/16/guest-post-18-dr-joanna-fadyl-seeing-the-changes-that-matter-qlr-focused-on-recovery-and-adaptation/

Zaitseva, E. (2018). Navigating the landscape of qualitative data in surveys with automated semantic analysis. *Big Qual Analysis Resource Hub.* http://bigqlr.ncrm.ac.uk/2018/12/11/guest-post19-dr-elena-zaitseva-navigating-the-landscape-of-qualitative-data-in-surveys-with-automated-semantic-analysis/

Case Study 6.1 Analysing Students' Employability Narratives: From Breadth of a Concept Map to Depth of Collective and Individual Narratives

Elena Zaitseva

This case study describes an approach to analysing a large volume of qualitative data generated as part of the UK's National Student Survey (NSS). The approach combines text mining with manual thematic analysis to examine career preparedness and employability discourses amongst final year students at Liverpool John Moores University [LJMU], UK. Our aim was to uncover trends and patterns in student feedback that could indicate how prepared students feel for their future after university, what sentiments are attached to the feedback, and where the university needs to focus its efforts to ensure the best possible start for graduates.

An average NSS institutional data set comprises between 80,000 and 100,000 words, amounting to approximately 200 pages of text (MS Word, Times Font 11). The analysis must be completed in a short time frame, allowing us to quickly unlock the main themes in student comments (open text responses), generate actionable insights and implement the changes during the course of the academic year. To handle this volume of material efficiently, the analysis was undertaken in three stages: (1) automated exploratory analysis; (2) concept seeding and analysis of a newly generated concept map, including sentiment exploration; and (3) manual, in-depth thematic analysis of the corpus.

The first two stages produce a 'bird's eye view' of the data, a 'panoramic scope of the project, which provides an overview of conceptual relationships' (Schostak & Schostak (2008, p. 208). **Automated exploratory analysis** was conducted using the text mining software Leximancer, which permits instant interaction with large volumes of qualitative data to reveal semantic characteristics of a text and patterns in the data (see also Chap. 5; see also Smith & Humphreys, 2006). Automatically discovered concepts, related to employability and career readiness, such as *work*, *placement*, *opportunities*, *practical* and *degree,* were explored, including their relevance, and positioning on the map.

The next stage, **concept seeding**, allows 'planting' user-defined concepts (other words frequently used by students such as *career*, *jobs*, *skills*, *professional*, *industry*, *future*, *applied*, *applying*, *confidence*, *graduate*, *opportunity* (singular), *placements* (plural), *practice*, which have not met the publication threshold set by the software) to produce a more tailored and informative concept map, allowing deeper, more granular exploration of semantic connections (Fig. 6.6). Sentiment analysis was enabled for identification, extraction and assigning weighted sentiment scores to the concepts.

The third stage involves looking at the data in depth and 'meaning making' by conducting a **manual thematic analysis** of the extracts that contributed to creation of concepts.

Narratives around concepts that had a high relevance score and high number of instances were explored first—to identify areas that were prioritised by students. The extracts were exported into MS Word for thematic analysis. Where extracts were abridged/truncated by the software, a search in the original data set was needed to extract the full comment.

In cases where contextual information was indicative of a subject specific experience, but lacking detail, the comment was located in the full data set to identify the subject area. This allowed us to explore/record the differences and subsequently make comparisons between the subject groupings. To improve the efficiency of the research, we decided to undertake a separate, comparative analysis of student data based on broad subject groupings in future.

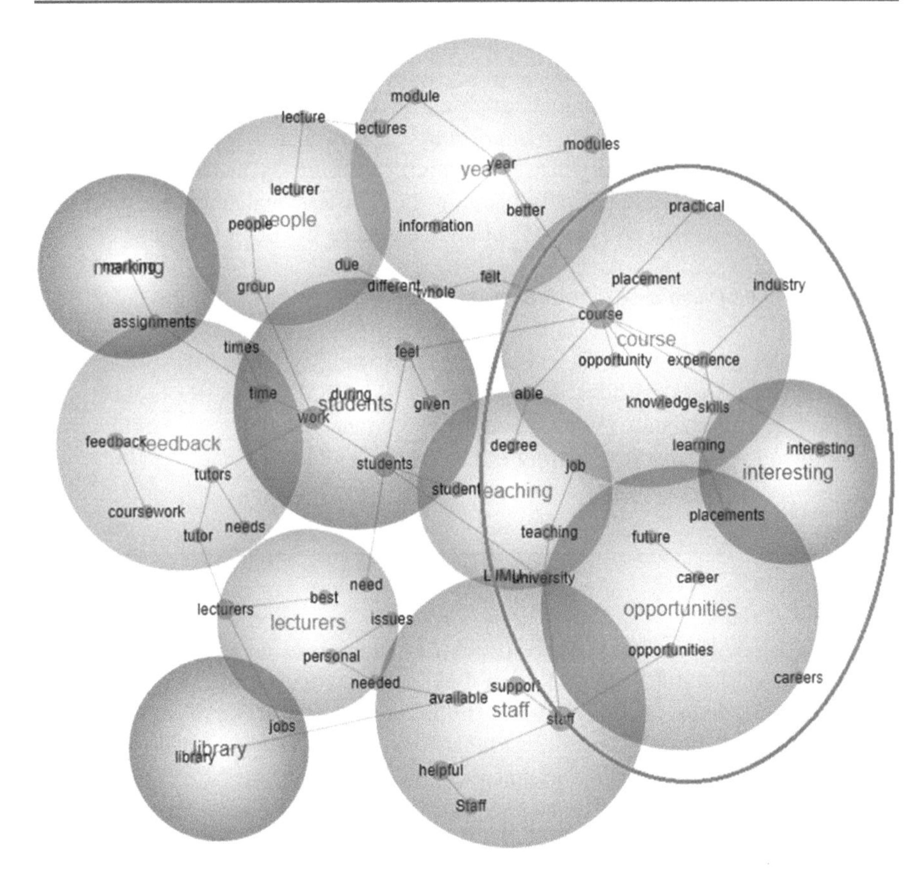

Fig. 6.6 Concept map with seeded concepts (employability related narrative is located within green circle)

Close attention was paid to the concepts that had a higher probability of being mentioned in unfavourable context. This was to identify areas that needed improvement, but also to verify decisions made by the software as even advanced natural language processing algorithms might lack accuracy in interpreting subtleties, contextual references, and colloquialisms expressed in comments. In some cases, going back to full comments in the original data set was needed to scrutinise the extracts, understand the validity of the sentiment, its strength and context.

Singular and plural forms of the words/concepts (e.g. *job* and *jobs, career* and *careers*) were analysed separately, as it gives an indication of a specific connotation or context in which students choose to use a particular word. These concepts/terms are not always located close to each other and are associated with a different contextual and often emotional background.

Exploring semantic connections depicted on the map via scrutinising direct quotes, allowed us to gain new insights into student experiences and priorities. For example, analysis of comments behind directly linked concepts such as *staff*, *opportunities*, *careers*, and *future* revealed how much students emphasised the high credibility of academic staff who are experts in their fields, and how their professional networks play an important role in students' ability to secure a work placement or future job.

You can read more about the outcomes of this analysis in:

Zaitseva E., & Finn C. (2021) Mining Employability Narratives—From Semantic Analysis to Institutional Strategy. In: E. Zaitseva, B. Tucker, and E. Santhanam (Eds.) *Analysing Student Feedback in Higher Education: Using Text-Mining to Interpret the Student Voice*. London: Routledge.

References

Smith, A. E., & Humphreys, M. S. (2006). Evaluation of unsupervised semantic mapping of natural language with Leximancer concept mapping. *Behavior Research Methods, 38*, 262–279. https://doi.org/10.3758/BF03192778

Schostak, J., & Schostak, J. (2008). *Radical research: Designing, developing and writing research to make a difference*. Routledge.

Zaitseva, E., & Finn, C. (2021). Mining employability narratives—From Semantic analysis to institutional strategy. In E. Zaitseva, B. Tucker, & E. Santhanam (Eds.), *Analysing student feedback in higher education: Using text-mining to interpret the student voice* (1st ed., pp. 133–145). Routledge. https://doi.org/10.4324/9781003138785

▶ Case Study 6.2 Young People and Food Choice in Transition

Mary Barker, Polly Hardy-Johnson, Sofia Strömmer, and Susie Weller

We employed the breadth-and-depth method in a piece of analytic work commissioned by the 'The Lancet' (Neufeld et al., 2022). Focussing on food choice in transition, the article brings together quantitative and qualitative studies exploring young people's autonomy and agency in a range of food environments in different global contexts. We were invited to conduct big qual analysis across data sets from ten qualitative studies undertaken in different sites in Bangladesh, Cote d'Ivoire, Ethiopia, India, Indonesia, South Africa, The Gambia, and the UK. Although the original studies were designed to address different research questions, they all had a similar broad substantive focus and employed in-depth interviews and/or focus group discussions to generate data.

To summarise steps one and two briefly, we conducted aerial surveying across the available data sets and pooled all relevant files. The resultant assemblage, comprising material from 650 participants, was organised into two groupings: young people and adults (including parents/carers and a range of other adults of salience in their lives). These two groups were each split further by how the local food environment was classed. This was defined by the original researchers and based on the High-Level Panel of Experts food system classification (HLPE, 2017). The three food environment types were: traditional (e.g. reliance on subsistence farming and markets), mixed (e.g. comprising both formal and informal markets, availability of inexpensive processed foods), and modern (e.g. mainly easily accessible formal markets, affordable staples and other foods). The HLPE classification includes emphases on the availability; affordability; promotion, and quality/ safety of food (HLPE, 2017). Generally, each of the original studies comprised one research site except for the multi-sited study undertaken in Bangladesh. Although we had been involved in some of the original studies (e.g. Barker et al., 2020, Weller et al., 2021) we were keen to involve, where possible, the original researchers and liaised with them over, for example, contextual material and methodological aspects. We then used the text mining software Leximancer (v. 4.51), which employs statistics-based algorithms, to conduct semantic analysis. Using word frequencies and co-occurrence we were able to identify and map key concepts and relationships and gain an overview of the nature and, to some extent, the content of the data in what felt like a vast and overwhelming assemblage. We now needed to move from exploring the breadth of the corpus to investigate the data in more depth.

For step three, we studied the resultant two-dimensional visual maps of key concepts and themes we had produced using Leximancer, and used these, along with data about the ranking of themes and the connections between concepts to run a series of Boolean queries. Some of these relationships were unsurprising, for instance, the connection between 'food' and 'eating' so we also focused on less salient themes or concepts that were present in some groupings but absent in others, for example, 'snacks' or 'street food'. Our 'test pits' were, therefore, the outcome of our Boolean queries and we exported the extracts from Leximancer to Microsoft Word and set about the task of undertaking a cursory reading of 500 pages of excerpts. Most of these were around 200 words in length. Some, often from focus groups were shorter, whilst the spread of text had to be increased around others to ensure we understood more about the context in which a response was given. We organised the extracts by the relationships between concepts and the age/food environment grouping. Each extract was examined briefly, with decisions made about those to select or dismiss reached collectively. Those dismissed included ambiguous or irrelevant statements. Whilst tempted to do so, we resisted reading the material in detail as we simply did not have the time or resources. We felt we had to balance this with dedicating enough time to ensure we understood what had been said and the context in which it was communicated in order to do justice to the data and to the investments made by the original researchers. Sorting and sifting through the extracts was a collective exercise conducted by a small team of researchers. This process enabled us to begin to explore and discuss similarities and differences between participant's responses across the different food environments, but it also pointed to areas where new Boolean queries needed to be conducted. We continually worked back and forth between the maps, the extracts, and the transcript data. We discussed and documented our decision-making, all the time refining our thinking. The selected extracts, organised into our age and food environment groupings, were then ready for more in-depth work using thematic analysis.

You can read more about the outcomes of this analysis in:

Neufeld, L. M., Andrade, E. B., Ballonoff Suleiman, A., Barker, M., Beal, T., Blum, L. S., Demmler, K. M., Dogra, S., Hardy-Johnson, P., Lahiri, A., Larson, N., Roberto, C. A., Rodríguez-Ramírez, S., Sethi, V., Shamah-Levy, T., Strömmer, S., Tumilowicz, A., Weller, S., & Zou, Z. (2022) Food choice in transition: adolescent autonomy, agency, and the food environment. *The Lancet, 399*, 185–197.

References

Barker, M., Hardy-Johnson, P., Weller, S., Haileamlak, A., Jarjou, L., Jesson, J., Krishnaveni, G. V., Kumaran, K., Leroy, V., Moore, S., Norris, S., Patil, S., Sahariah, S., Ward, K., Yajnik, C., Fall, C., & the TALENT collaboration. (2020). How do we improve adolescent diet and physical activity in India and sub-Saharan Africa? Findings from the Transforming Adolescent Lives through Nutrition (TALENT) consortium. *Public Health Nutrition*. Online.

HLPE. (2017). *Nutrition and food systems*. A report by the High-Level Panel of Experts on Food Security and Nutrition of the Committee on World Food Security. Rome.

Weller, S., Hardy-Johnson, P., Strömmer, S., Fall, C. H. D., Banavali, U., Chopra, H., Janha, R., Joseph, S., Joshi Reddy, K., Mengistie, M. A., Wrottesley, S., Kouakou, E., Barker, M., & the TALENT collaboration. (2020). 'I should be disease free, healthy and be happy in whatever I do': a cross-country analysis of drivers of adolescent diet and physical activity in different low- and middle-income contexts. *Public Health Nutrition*. Online.

Case Study 6.3 Moving Home Across the Life-Course

Rosalind Edwards, Susie Weller, Lynn Jamieson, and Emma Davidson

Although the 'Working across qualitative longitudinal studies' project, outlined in Chap. 2, was primarily a methodological endeavour, care and intimacy formed our substantive interest and we used this focus as a means of testing out early versions of the method as our ideas developed. This example details some exploratory work we conducted across an assemblage of data sets housed in the Timescapes Qualitative Longitudinal Data Archive (https://timescapes-archive.leeds.ac.uk/). For step two, we used the web-enabled text mining software Leximancer (v4.5) to produce two-dimensional concept maps detailing the main themes and concepts across the corpus (see also Heath & Swabey, 2014). The software uses word frequencies and co-occurrence to create clusters of terms that are inclined to feature together in a text. Exploring how words relate to one another, it is premised on the idea that the terms surrounding a given word determine its meaning. Accordingly, it mines thousands of iterations to see how these words 'travel together'. We found it to be a very useful tool in that it is relatively easy to learn and it processes a large amount of data in a short space of time. Figure 6.7 is an example of the output from step two.

Our approach was exploratory and aside from having a broad interest in vocabularies and practices of care and intimacy over time, we were

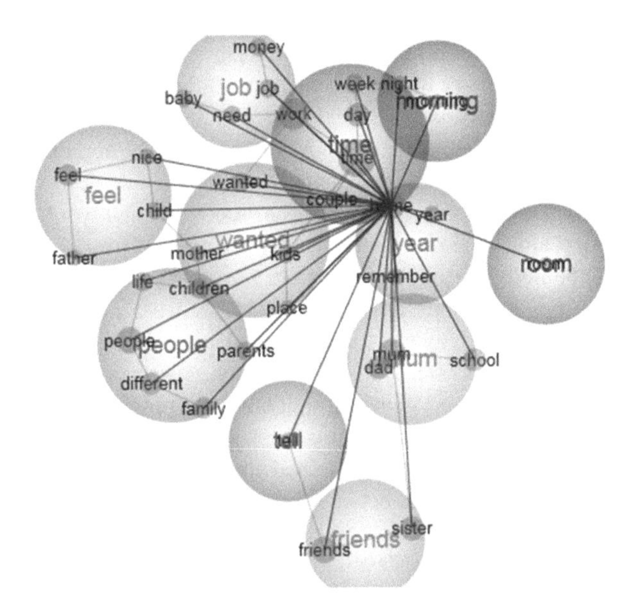

Fig. 6.7 Example concept map generated using Leximancer

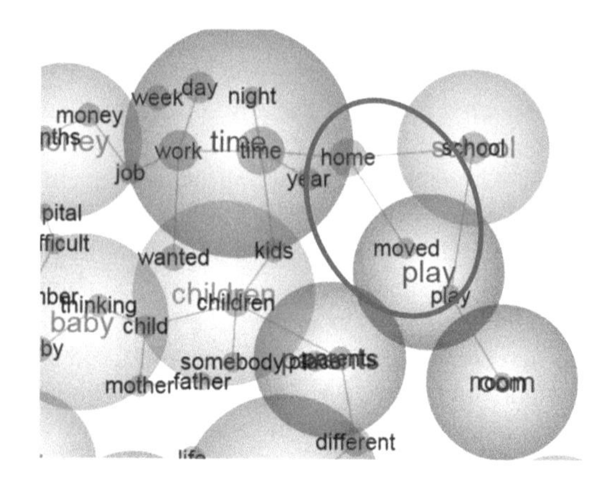

Fig. 6.8 The relationship between 'home' and 'move/d'

keen to identify areas of interest that had not featured as key foci in the original Timescapes studies. We organised our assemblage into gender and generation cohort groupings. Taking the composite data set in its entirety 'house/home' was the first theme that did not feature as an explicit focus of Timescapes. On closer inspection of the individual maps for each grouping, the relationship between 'home/house' and 'moving/ed' seemed particularly salient for all women, regardless of cohort-generation (Fig. 6.8). This was not the case for men.

We were curious about this difference and wondered what variations there might be across women from different cohorts. When considering the gender cohorts separately the connection seemed more salient for women born before 1950. We decided to dig test pits in this area and used the query function to conduct Boolean searches around 'home' and 'moved'. The query for 'all women' and 'all men' returned 203 and 161 matches respectively.

Opening the extracts revealed more detail about the context. Akin to other text mining software, Leximancer's interface enables extracts to be viewed in context with ease. We could see very quickly whether the results of a search were relevant. All but a few extracts from the Boolean search of 'home/house' and 'moving/ed' returned data of interest.

As Fig. 6.9 highlights although the file is now part of one data assemblage, it is still linked to its original project. We exported the selected extracts into Microsoft Word and organised the 136 pages of excerpts into gender and generation-cohort groupings. Informed by a review of the literature on moving home we adopted a theoretical approach to sampling that combined our emphasis on exploring similarities and differences between/across cohort generations with a focus on those 'left behind' in home moves. Our main rationale for this area of substantive interest was that life course approaches have tended to document the lives of those moving home, at the expense of those left

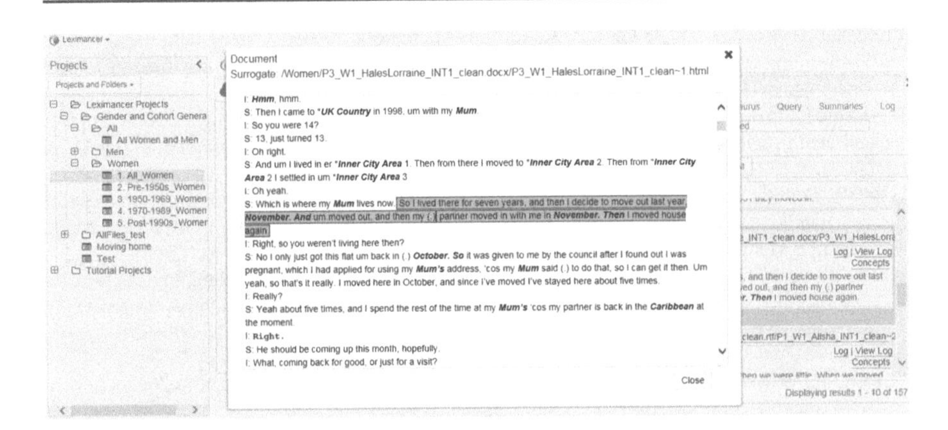

Fig. 6.9 Viewing extracts-in-context using Leximancer

behind in the process. We used our data assemblage to understand the implications of/for those left behind in relation to caring and intimate relationships, identities, and practices.

For the final phase of step three, we sifted through the extracts conducting two further cursory readings. The first focused on identifying stories of those 'left behind' in home moves, whilst the second was concerned with categorising the cases and identifying potential short- and long-case studies—with the former referring to isolated short stories about being 'left behind' and those where multiple extracts had been returned for the same case suggesting that being 'left behind' may have featured as a more dominant narrative in their interviews(s). The short case studies were generated from reading a little wider than the test pits (with one exception where the narrative was so complex that further reading was required). The long case studies were constructed with the 'test pit' extracts at the heart. These cases were taken forward for more in-depth interpretative work in step four.

You can read more about the outcomes of this analysis in:

Edwards, R., Weller, S., Davidson, E., & Jamieson, L. (2021). Small Stories of Home Moves: A Gendered and Generational Breadth-and-Depth Investigation. *Sociological Research Online*. https://doi.org/10.1177/13607804211042033

References

Baker, S. E., & Edwards, R. (2012). *How many qualitative interviews is enough?* Discussion Paper. NCRM, Southampton. http://eprints.ncrm.ac.uk/2273/

Bruns, A. (2013). *Faster than the speed of print: Reconciling 'big data' social media analysis and academic scholarship*. *First Monday, 18*.

Cretchley, J., Rooney, D., & Gallois, C. (2010). Mapping a 40-year history with Leximancer: Themes and concepts in the journal of cross-cultural psychology. *Journal of Cross-Cultural Psychology, 41*(3), 318–328.

Davidson, E., Edwards, R., Jamieson, L., & Weller, S. (2019). Big data, qualitative style: A breadth-and-depth method for working with large amounts of secondary qualitative data. *Quality & Quantity, 53*(1), 363–376.

Gapp, R., Stewart, H., Harwood, I., & Woods, P. (2013). *Discovering the value in using Leximancer for complex qualitative data analysis*. British Academy of Management Conference, Track: Research Methodology—Workshop.

Heath, A., & Swabey, K. (2014). Content and Leximancer analysis as methodological tools to study formal school grievances: An archival study. In *2014 joint Australian Association for Research in Education and New Zealand Association for Research in Education Conference* (pp. 1–5).

Lewins, A., & Silver, C. (2020). *Leximancer v5.0: Distinguishing features* https://www.surrey.ac.uk/sites/default/files/2020-12/cnp-leximancer-5-review.pdf

Nelson, L. (2021). Cycles of conflict, a century of continuity: The impact of persistent place-based political logics on social movement strategy. *American Journal of Sociology, 127*(1), 1–59.

Neufeld, L. M., Andrade, E. B., Ballonoff Suleiman, A., Barker, M., Beal, T., Blum, L. S., Demmler, K. M., Dogra, S., Hardy-Johnson, P., Lahiri, A., Larson, N., Roberto, C. A., Rodríguez-Ramírez, S., Sethi, V., Shamah-Levy, T., Strömmer, S., Tumilowicz, A., Weller, S., & Zou, Z. (2022). Food choice in transition: Adolescent autonomy, agency, and the food environment. *The Lancet, 399*, 185–197.

Richards, T. (2008). Survey strategies in landscape archaeology. In B. David & J. Thomas (Eds.), *Handbook of landscape archaeology* (1st ed.). Routledge.

Zaitseva, E., & Finn, C. (2021). Mining employability narratives—From semantic analysis to institutional strategy. In E. Zaitseva, B. Tucker, & E. Santhanam (Eds.), *Analysing student feedback in higher education: Using text-mining to interpret the student voice*. Routledge.

'Deep Excavations': In-Depth Interpretive Analysis

7

7.1 Introduction

This chapter considers the fourth step in our breadth-and depth method of working with large amounts of qualitative material, taking us fully into the depth element. In the previous chapter we covered a critical turning point stage in our analytic method, shifting from the breadth of data assemblage into the depth of digging into the detail of the data, akin to archaeological test pit sampling. We are now building on the step three examination of multiple small extracts of data in the sampled test pits to undertake engagement with whole cases—a process of deep excavations in terms of our archaeological metaphor. This is a movement into a greater depth of qualitative analysis with the ultimate purpose of bringing the resultant rich analysis back into conversation with the breadth analyses of previous steps. In step four of the breadth-and-depth analytic method we move away from working with 'bags of words' as we did in step two and from small sampled test extracts as in step three, to return to the material comprising the original big qual corpus and pull out selected whole cases.

In this chapter, we will:

- explain what we mean by our use of the term 'cases' for the purposes of step four and how these are selected for deep excavation
- cover the central tenets of qualitative analysis and the strengths it can bring to understanding the various textual forms of data that the breadth-and-depth analytic method largely works with
- detail several different possible qualitative analytic approaches that can be adopted for the in-depth qualitative analysis and practicalities
- consider how various forms of analysis can be used in combination to complement or contrast analytic insights
- reflect on the importance of bringing the interpretive analysis of step four back into engagement with the more structural understandings of data generated in the 'geophysical' step two and 'shallow test pit' step three

© The Author(s), under exclusive license to Springer Nature Switzerland AG 2023

S. Weller et al., *Big Qual*, https://doi.org/10.1007/978-3-031-36324-5_7

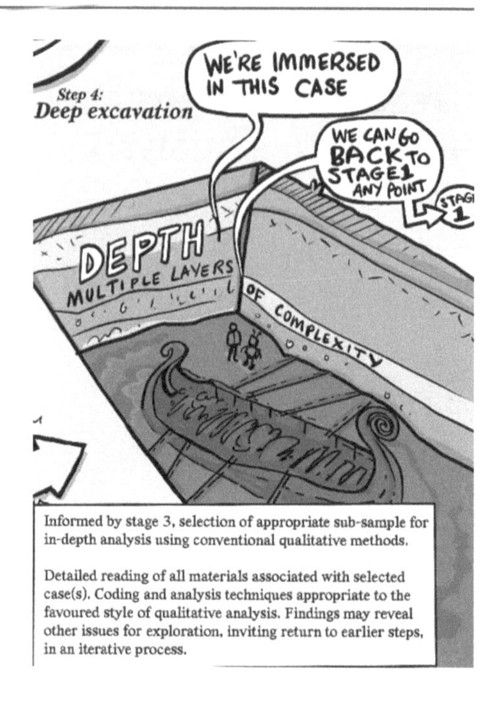

Fig. 7.1 Archaeological metaphor and step four of the breadth-and-depth method. (Illustration created by Chris Shipton https://www.chrisshipton.co.uk/)

7.2 Metaphorical Foundations

We draw on the archaeological metaphor of deep excavation to characterise step four of our breadth-and-depth method (Fig. 7.1). As we have argued in Chap. 2, this metaphor offers a framework of orientation, to understand our breadth-and-depth corpus as a landscape containing interesting features to be dug into. It conveys the complex reality of stages in how researchers can deal with (relatively) large amounts of qualitative data while retaining the integrity of a contextual and detailed qualitative approach. Deep excavation captures how we delve below the surface and into the detail of some cases informed by the previous geophysical survey and shallow test pits. As with archaeological excavation (Drewett, 1999), the aim of in-depth qualitative analysis is to uncover, identify, and interpret in context. Also akin to archaeology, in-depth qualitative analysis requires thought, time, and flexibility, and is shaped by the theoretical framework and techniques adopted, and the pragmatics of the nature of the 'site'/data. Unlike archaeology, however, the site is not destroyed through the excavation; data and cases can be revisited and dug into again and again, as with working with secondary archived qualitative data.

7.3 Cases for Deep Excavation

In moving from examining extracts of qualitative data from the test pit samples into an engaged analysis of selected whole cases pulled out from the original big qual corpus, we need to consider what we mean by a 'case' for step four. Cases are units of analysis that enable a close, detailed, and contextualised exploration of the complexity of the substantive topic that you are investigating. What constitutes a case lies in the meeting point of the 'internal' intellectual purpose of your project and the 'external' parameters of the various sets of qualitative material that went into creating your merged corpus (Emmel, 2013). When you are working across multiple small-scale data sets, this can raise knotty questions about what to take as a case or unit of analysis for the deep excavation elements of the breadth-and-depth method. Different data sets may speak to the same set of intellectual and substantive puzzles but they may have generated material from a variety of sources in different forms. This is an issue that you will already be aware of through the step one aerial survey review of the data sets (see Chap. 4). These differences in data sources across a big qual corpus mean that you may need to consider whether a case is, for example, an individual; a set of participants such as a sibling group or members of a messaging app group; the research encounter itself as in interviewer–interviewee verbal or textual exchanges; a single or multiple set of organisation/s or institution/s such as a charity or several schools; or a period of time or a geographical location. In the *Possibilities of Paradata* case study (Case Study 7.1), the unit of analysis was documents in the form of survey booklets, while in the *Growing Up on the Streets* case study (Case Study 7.2) it was researcher ethnographic reports, and in the *QUALIDEM* case study (Case Study 7.3) the unit was interviews with individual citizens.

Once you have a firm idea of what constitutes a unit of analysis for the purposes of your project, the guide for your decisions about selecting cases rests on an interaction between the relationship you envisage between theory and data as deductive, inductive, abductive, or retroductive (see Chap. 2); your preliminary keywords and themes searches and analyses; your conceptual approach; and your substantive research topic and questions. Looking at your previous step three analysis in tandem with the metadata that you have available for the units in the overall research study, the particular cases that you decide to select for a deep excavation are chosen as fruitful to pursue because they hold the promise of insight into the specific features of the social processes and meanings that are the concern of your big qual project. Informed by the relationship between theory and data you envisage, cases with potential are not just those that are typical of recurrent and common issues from the step three computer-aided preliminary thematic analysis of extracts. They can also be those that seem especially revealing of a set of understandings and circumstances in some way, and/or in the light of having a comparative substantive focus and questions for your research, the cases may be selected to form a strategic comparison with another or other selected case/s. Cases may also be identified because they stand out as unusual. Being unusual can mean that the test pit analysis has indicated a different aspect of the topic to other cases—an uncommon presence. Or it could mean that a recurrent theme in other cases simply is not present in the case—a curious absence perhaps. The *QUALIDEM* case study (Case Study 7.3) describes the stages of the research team's 'big abductive' approach to their merged secondary qualitative data sets to build theory looking specifically for anomalies and absences.

Looking at the factors in play in the selection of cases, for example our own research is conducted from an interpretive epistemological perspective, which is a broad range of understandings that are loosely concerned with understanding social phenomena from the perspectives of those involved. Hence, interpretive analysis takes the form of explanations of how others perceive and make sense of their day-to-day life and interactions (e.g. Rosenthal, 2018; Yanow & Schartz-Shea, 2006). Our broad substantive interest was in comparisons of care and intimacy by gender and age cohort, informed by theoretical ideas about shifting vocabularies and practices over time (e.g. Giddens, 1992). This led us to focus on individuals and their interview transcripts as our cases. Our step two breadth keyword analysis had shown us that co-occurrence of 'home/house' and 'moved/ing' was strongest among the women participants in our data corpus, and that generally the older the generation, the more home/house and moved/ing were salient for their accounts, with the exception of men in their 20s through to early 40s. This indicated, for instance, women as potential typical case units, and potential puzzling cases involving 20–40-year-old men. The shallow test pit analysis of interview extracts containing co-occurrence of 'home/house' and 'moved/ing' for step three generated recurrent themes of passivity or activity, moral or taken-for-granted stances, home as people as well as place, and anxieties. Passivity was not so prevalent in men's accounts and anxieties were voiced more by women, so again this helped to indicate selection of cases. We also glimpsed the possibility of different types of stories in contexts where interviewees had either moved home themselves or had stayed put while others moved away from or closer to them—another guide in our selection of cases for in-depth qualitative analysis from our step three test pit analysis of our corpus (see Edwards et al., 2021).

7.4 Qualitative Analysis

The sort of interpretive qualitative analysis that is used for the deep excavation case-level analysis of step four is a process that the majority of qualitative researchers will recognise, value, and be comfortable with. It is an immersion in data at a scale that qualitative researchers feel uses the strengths of qualitative analysis, that is, it is detailed, and sensitive to changing context and multi-layered complexity. Equally though, the central tenet of interpretive analytic practice may be one that researchers more at ease with quantitative work and step two of the breadth-and-depth method may find rather nebulous. In-depth qualitative analysis focuses on rich detail to represent intricate social realities and produce nuanced social explanations. Hence, this step will also involve placing the shallow test pit extracts of step three back into their context as a whole. Deep excavation as part of the archaeological process of working with big qual brings depth back into conversation with breadth.

Qualitative methods of analysis are concerned with transforming and interpreting data to identify and understand the complexities and processes of the social world. The defining feature of qualitative material, and its in-depth interpretive analysis, is its attention to understandings of the whats, whens, whys, whos, and hows. The qualitative data that form case studies can take a textual, visual, and/or audio form. Texts in the form of research interview transcripts underpinned the development of the breadth-and-depth method, but the method is appropriate for large amounts of qualitative data taking other textual forms, such as open-ended responses to survey questions (e.g. Elliott, 2008), research interviews conducted via email (e.g. Fritz & Vandermause, 2017), blogs (e.g. Hookway, 2008), or audio or written solicited or unsolicited diaries (e.g. Bartlett, 2012). Qualitative interviews, whatever the variation in form, style, or tradition, and in the transcription of interviews that represent them (see Chap. 1), involve interactional exchange of dialogue between two or more participants, in-person or in other remote and possibly even asynchronous contexts. They also involve coverage of topics, themes, or issues within a fluid and flexible structure, and an understanding of the relevance of context (Mason, 2002). Working with in-depth analysis of whole cases, the 'data' to be subject to deep excavation are taken to be every word said in an interview or written in a document, informed by awareness of the context of the interview and of the interviewees' situation or the circumstances of text production. Through detailed qualitative analysis we can explore the texture and weave of everyday life; the understandings, experiences, and imaginings of our research participants; how social processes and relationships work; and the significance of meanings (Mason, 2002).

What constitutes qualitative data analysis is subject to debate (Coffey & Atkinson, 1996). Nonetheless, broadly it concerns making sense of material that is derived from qualitative methods of data generation. It involves some systematic form of breaking down and organising data (in this case) in meaningful ways so as to address your research aims and answer your research questions. It typically involves, variously:

- Reading, re-reading, and reflecting on your material
- Searching for, identifying, and describing patterns of meaning and processes
- Comparing and contrasting meanings and processes
- Searching for relationships between meanings and processes

An in-depth analytic engagement with a transcript or text frequently starts with a detailed reading, looking for narratives or for themes, meanings, and processes, paying attention to the contexts of utterances. This often includes attending to the interactive dynamic—what is prompted and what is said spontaneously. It also typically includes looking across all that an interviewee or writer communicates to explore whether what is expressed in one place is complemented, elaborated, or contradicted by what is said in others. For most qualitative researchers who are undertaking deep engagement with cases, to take extracts from material without awareness of the whole data unit and set within which it is embedded, and without attention to 'context', is not best practice for in-depth analysis.

Qualitative researchers often produce summary pen portraits of their research units, comprising a brief history of each case, condensing, and highlighting key topics and circumstances to help them keep track of cases in a sample, or perhaps more extensive case profiles or histories providing thicker description (Neale, 2016). For

4. LOUISE

Bionote: Louise, born in 1989, was a White British woman who lived in suburban SE England with her parents and two co-resident siblings. She has an older sister who lives independently. By Wave 3, she was living with her boyfriend's family on a part-time basis. She left school at 16 to become a receptionist (she also engaged in a second job and was planning to do voluntary work overseas). Her mother's occupation was a teaching assistant, whilst her father worked in a leasing company. She was interviewed on three occasions in 2003, 2007 and 2009. [Nb. (pseudo)surname was Kirk].

Test pits: 3 in total comprising: 1 x Wave 1 and 2 x Wave 3.

Mini case: Louise had lived in the same family home for the majority of her life '*I was about one when we moved here*'. By her third interview she described splitting her time between living in her parental home and that of her boyfriend's family. She said '*I like to have my own space. That's why even though I spend all my time at my boyfriend's home when I come back here it's like, 'Oh at least I have home''*. For Louise, having an alternative home and her own separate space to which she could return was important. For her, this frequent movement, back and forth between homes was an active choice; an attempt to balance difference aspects of their lives.

Fig. 7.2 Example of a case note for the home moves step four breadth-and-depth analysis

our breadth-and-depth research on home moves, as part of our reviewing of data and to support our searching and comparing, we produced contextual notes for each case covering the personal characteristics and circumstances of the research participant concerned, a summary of the contents of their interviews, and also noting the original project that the case resided within and any field notes provided by the original researchers to enhance our understanding (see Fig. 7.2). Our process here speaks to the importance of attending to the context in which data are produced rather than seeing data as if autonomous of or disengaged from their origins.

7.5 Forms of In-Depth Interpretive Analysis

There is no particular or perfect technique or combination of approaches for in-depth, interpretive qualitative case analysis that should be used for the deep excavations phase of the breadth-and-depth method. There are a diversity of qualitative analytic strategies and in-depth techniques that might be applied, which illuminate, variously, social meanings, subjectivities, activities, processes, constructions, and discourses (e.g. Coffey & Atkinson, 1996; Grbich, 2007). In other words, there are different articulations of what constitutes knowledge and understanding that are pursued in different types of analytic approaches. Nonetheless, whatever the possible approach to in-depth analysis, the various strategies share the purpose of working with and interpreting qualitative material so as to identify, describe, and explain the complexities of the social world.

Some—certainly not all—of the most well known of the methods for analysing written texts are presented here to indicate the various analytic practices that may be applied to your selected cases in step four. Which in-depth analytic technique you adopt depends on your epistemological stance (your philosophy of what constitutes knowledge of the social world), your theoretical approach (your abstract idea about how the social world works), the concepts you have adopted (applied features of your theory), the substantive focus of your study (your research topic), and finally the practicalities of your case study (such as the interview transcripts, open survey responses, diaries etc. we mention above).

Thematic analysis (e.g. Braun & Clarke, 2006, 2020). This is a very widely used set of analytic tools where you identify, analyse, and report patterns and recurrent themes that you see in your data. It can have various inductive and deductive underpinnings. We have already used preliminary versions of coding-driven and reflexive thematic analyses in steps two and three for our geophysical keywords and test pit sampling across the merged corpus. But, as we illustrate later on in this chapter, the analytic method may also be applied more deeply to the cases chosen for step four, searching thematically within selected units of analysis rather than across a corpus, to draw out and interpret richness and complexity. Themes can be manifest or latent, broad categories or repeated phrases, identified deductively and/or inductively. Thematic analysis usually involves segmenting and classifying parts of the interview transcript, and then linking and refining related categories. Tied to thematic analysis is ***framework analysis*** (Ritchie & Lewis, 2003), which involves organising your data to identify commonalities and differences across cases as well as within individual cases. It provides a systematic structure for your analysis around the construction of a matrix of cases and themes.

Grounded analysis (e.g. Charmaz, 2014; Flick, 2018). Grounded analysis is strongly related to the formal grounded theory approach but is concerned specifically with the data analytic element of the methodology and tends to be used in a looser fashion than the various versions of the theory. Grounded analysis is where you induct meaning from the specifics of your data in as open a way as possible to start with. You then move from this specific induction into an iterative process of categorising, selective coding, and conceptualisation. The idea is that your knowledge about and theorisation of the social topic emerges from the data in a grounded fashion rather than being imposed on it. Applying grounded analysis to case study interview transcripts or written texts as part of step four of the breadth-and-depth method would be a way of understanding and abstracting how social processes work.

Narrative analysis (e.g. Andrews et al., 2013; Reissman, 1993). This analytic technique focuses on spoken or written stories told by participants to illustrate or account for aspects of their experiences and actions, and which have significance for themselves and their audience. Narrative analysis is based on the idea that much of our communication with each other is through stories that convey our understandings and behaviour. Participants' stories have a beginning, a middle, and an end, and they may fall into a particular genre (e.g. see the *Possibilities of Paradata* Case Study 7.1). Using narrative analysis, you look at how research participants construct and sequence the stories they tell in the interviews to throw light on their experiences, interpretations, and rationales. The aim of using narrative analysis for the step four case studies would be to reconstruct the holistic meaning of the stories that the research participants in the selected cases have told about themselves in order to access the underlying order of their accounts of the topic under investigation.

Conversation analysis (e.g. Liddicoat, 2011; Sidnell, 2010). You use this method of analysis to explore the procedures that speakers use to communicate. The assumption is that social structures are reflected and social order is achieved in the detail of social encounters and interactive communication. The analysis is fine-grained, to focus on forms of exchange in the conversation, such as sequencing and turn-taking, expressive niceties and silences, assertions and declarations, turns of phrase and repair mechanisms, and the functions they serve. Where your cases take the form of interview transcripts or written exchanges, then the focus of your analysis will likely be on the minutiae of how research interviews are achieved in various circumstances.

Discourse analysis (e.g. Fairclough, 2010; Taylor, 2013). Discourse analysis is a set of techniques for identifying sets of ways of thinking about, of knowing and being able to write or speak about, a situation or behaviour or group of people, to map which of these are dominant and powerful and which are marginalised. There is also the identification of what is not said, that is, silences on a topic or aspect of a topic. The analytic focus is on discourse itself, as strategic textual and verbal devices to establish one version of the social world as against other possible versions, and thus constituting how we perceive what is going on in the social world. Discourse is regarded as working to achieve particular ends, and discourse analysis as searching for this purpose. For transcript or text-based cases, discourse analysis can be used to address how, through language, aspects of society and culture are spoken or written into being.

The above is just a small selection of the extensive variety of analytic methods that can be applied to qualitative research material. It is also a brief skim over the top of the methods that we have described, which are all more complex and have distinct internal variations. Which analytic technique or combination of techniques you adopt for your in-depth qualitative analysis of your selected cases is determined by your philosophical and conceptual approaches, the theoretical and substantive orientation of your research project, and the opportunities and constraints of the 'external parameters' of data that you are working with, for example transcribed from spoken or written material, or involving an individual or a group.

In the breadth-and-depth analysis focusing on young people's food choices conducted by one of us as part of another team of researchers (Neufeld et al., 2022), which pooled 11 data sets from 10 different studies involving 652 participants overall, step four used thematic analysis of selected cases to explore the influence of different sorts of food environment (see Case Study 6.2). The team coded material that discussed topics pertinent to food choice and food environment and organised them under emergent descriptive themes. For example, talking about eating out with friends was coded as 'food as social currency'. Such broad codes were then refined by further analysis of the relevant data to identify the specifics of that theme, from which commonalities and differences between food environments could be identified.

A different analytic method example is provided by our home moves research. In this, we conducted a 'small story' form of narrative analysis in step four, which provided us with fine-grained micro-analysis of fleeting aspects of lived experience in time and space. Just to give a taste of the narratives, we identified a typology of three main forms of stories about home moves: tales of staying put, where the participant describes being and feeling left behind, often conveying anxiety or ambivalence or, less often, being proactive in others' moves; return histories comprising accounts of revisiting or returning to people or places as home; and pendulum plots, which are characterised by repeated or rhythmic return to people and places of former residence, occurring with high frequency over a short period or over a much longer period of time. This project involved one form of analysis, focusing on participants' narratives, for the deep excavations of step four. While some accounts compare different analytic strategies in an effort to help researchers choose which technique is the best fit to apply in their studies (e.g. Braun & Clarke, 2020), others have advocated combining different forms of in-depth qualitative analysis to provide complementary insights.

7.6 Mixing and Matching Interpretive Analyses

Your big qual corpus, made up as it is of material from merged data sets, may comprise original studies that each emerged from different disciplines or approaches, even where they have a shared substantive focus. For example, our own Timescapes corpus combined oral history, sociological, psycho-social, and social-psychological approaches (see Chap. 2). The advantages of bringing together data from different qualitative sources is already advocated in discussions of research synthesis and systematic reviews (e.g. Campbell et al., 2006; Sandelowski & Barroso, 2006), such as to promote the value of qualitative research and support understanding of evidence bases. But there are also insights to be gained from applying different methods of in-depth qualitative analysis to the same cases.

There are methods of analysing qualitative material in depth that take the form of several phases involving different ways of working through the data towards a comprehensive analysis, and which lend themselves to exploration of case studies. For example, the hermeneutic reconstructive method seeks to reconstruct the rules by which an interviewee has organised their telling of events and experiences, in the context of social constraints and options. The analytic phases consist of constructing a chronological sequence of events; then working systematically through passages of interview transcript text building up hypotheses for the nature of the stories being told, confirming or rejecting these hypotheses along the way; and finally bringing the surviving overarching hypothesis alongside the sequential event analysis (e.g. Rosenthal, 1993). Another multi-stage method is the Voice-Centred Relational Method or Listening Guide (e.g. Gilligan et al., 2003), which seeks to listen to and trace what is regarded as the inner, authentic voice of the research participant. The Method/Guide involves four main, sequential readings of an interview transcript for: (1) the 'plot' of the interview and your reaction to it; (2) how the participant speaks about themselves; (3) relationships with other people; and (4) cultural contexts and social structures.

The forms of application of these multi-stage deep analysis techniques to big qual data are dependent on the amount of data and the size of the research team involved (see Chap. 1). The breadth-and-depth method was developed to analyse an extensive qualitative corpus (in our case just under 400 interview transcripts) by sole researchers or small research teams. Working as sole researchers on their respective projects means that Mauthner and Doucet (1998) recommend the Voice-Centred Relational Method (also known as the Listening Guide) as best suited to a select number of cases from data sets of around 40 interviews, 'tuning the ear' to issues before working with other transcripts. This is, in effect, working in the opposite direction to the breadth-and-depth method—starting from intensive excavation (step four), but then remaining at the equivalent level of shallow test pits (step three). In contrast, Fontaine and colleagues, forming a research team of four, adapted the Voice-Centred Relational Method to analyse a big qualitative documentary data set of under 400 short narrative extracts from clinical case notes held by a social service intervention—or 'creatively mine' the case notes in their terms (2020, p. 9). After conducting an initial thematic analysis phase somewhat akin to a combination of steps two and three of the breadth-and-depth method using the software package Dedoose, the research team then conducted readings for, respectively, structure and plot, named subjects, and core stories of what was said as well as left unsaid, as would be the practice in undertaking step four of the breadth-and-depth method. Further contrast to these examples is provided by research teams working with whole data sets and cases along lines akin to the breadth-and-depth ordering of analytic steps. The *QUALIDEM* case study (Case Study 7.3) multi-phase abductive coding approach worked with the corpus and then theoretically anomalous cases before re-engaging with the corpus, and the *Growing Up on the Streets* case study (Case Study 7.2) worked with deductive capabilities coding of the corpus and longitudinal case histories from selected participants.

Coffey and Atkinson (1996) call for recognition of the value of multiple analytic strategies to look at the same data set from contrasting angles. They encourage researchers to try out different analytic approaches on their material in order to elaborate rich and variegated alternative aspects of social reality. While they warn against thoughtless mixing of many analytic approaches that may not be compatible, cautioning that the range of qualitative analytic techniques cannot be combined at will, and against simplistic attempts at methodological triangulation, they state that:

> There is much to be gained from trying out different analytic angles on one's data. New insights can be generated, and one can sometimes escape from analytic perspectives that have become stereotyped and stale. We therefore want to encourage a (modestly) playful approach to the diversity of research approaches .... We can use different analytic strategies in order to explore different facets of our data, explore different kinds of order in them and construct different versions of the social world ... The more we examine our data from different viewpoints, the more we may reveal—or indeed construct—their complexity. (1996, pp. 13–14)

Coffey and Atkinson's own demonstration of the value of complementary analytic strategies is through a case study, where they pull out a small corpus of interviews from another, larger project about social science doctoral strategies, to focus on social anthropology students and academics. Their combination of thematic, narrative, and semiotic analyses illuminates how early career and experienced scholars viewed key elements of becoming and being an anthropologist. In a similar vein, but using interviews over a period of time with a single research participant as a case study rather than working across cases, two of us (Edwards & Weller, 2012) explored the perspectives on aspects of our interviewees' subjectivity. We discuss the application and insights from both a thematic analysis and an I-poem analysis that comprises the second reading in the Voice-Centred Relational Method—and the differential positioning of the researcher analyst through the process. We argue that these distinct insight angles are complementary rather than contrasting or competing.

7.7 Bringing Depth Back into Conversation with Breadth

From the point at which we produced our merged corpus as an outcome of the aerial survey step of the breadth-and-depth method step one (Chap. 4), our analysis has been working from the breadth of geophysical analysis in step two to identify keyness (Chap. 5) into the depth of test pit analysis of step three (Chap. 6) and now, in this chapter, the deep exploration of in-depth interpretive case analyses. This is not the end of the process, however. The breadth-and-depth method is an iterative technique.

Each of the analyses from steps three and four may lead us back to revisit any of the previous steps for further scrutiny. The in-depth analyses from thematic, narrative, and/or discourse techniques and so on in the deep excavation that we have been describing here may generate new ideas and issues. These insights may mean that, for example, you look again at test pits that you did not pursue as cases for step four and select some further examples as cases, or that you now identify as pertinent preliminary themes in some transcript extracts that you did not spot originally as being of any interest. Or you may return to step two to rethink the results from your geophysics keyword and keyness analyses, perhaps running a search for a particular word or set of words in the light of the new insights from your deep excavations, and then working your way through steps three and four once more on the basis of the new geophysics. You may even go back to step one, having realised that some archived small-scale qualitative data sets that you had not thought appropriate for your purposes on your first sweep because of their topic, focus, nature, or approach now may have something useful to tell you. This flexibility to move back and forth between steps in the breadth-and-depth method of analysis means that any significant issues that may not be spotted in step two's keyness analysis or step three's semantic mapping can be picked up in step four, and then traced back from depth through to breadth.

Whether you return to revise and repeat steps or not, once you have worked through to a finalised version of the in-depth analysis of whole cases for step four, the breadth-and-depth method remains iterative. The depth understandings generated from the deep excavations are then brought into engagement with and placed in the context of the breadth, more structural features identified in previous geophysics and test pit steps, so that the elements in the whole process inform each other. For instance, earlier we mentioned our application of the breadth-and-depth method for our project on 'home moves'. Our 'small stories' narrative analysis highlighted the nuances of family and personal relational interdependencies in home moves. Brought back into engagement with the key preliminary thematic elements identified for the corpus as a whole in steps two and three, we had a broader view of how these were shaped structurally by gender and generation. In turn, we unearthed how, for example, employment, finances, and residential tenure all provided conditions shaping the ability to do the right thing and feelings of anxiety in the considerations around breadwinning and home moves by fathers—our standout cohort of 20–40-year-old men in the co-occurrence of 'home/house' and 'moved/ing' in the corpus of accounts and illustrated in the gendered themes in the case fathers' somewhat nervous small stories. The gendered experience and telling of home moves was a significant feature made apparent through our breadth-and-depth approach, which allowed us to identify and compare themes across a large amount of qualitative data and then pursue them in depth through case analysis. Working with multiple merged data sets meant that we could conduct secondary data analysis across disparate qualitative studies, making use of possibilities of strategic comparisons that cannot be addressed in a discrete small-scale study.

▶ Case Study 7.1 The Possibilities of Paradata*: An Historically Situated Exploration of 'BigQual' Marginalia

Ann Phoenix, Heather Elliott, Janet Boddy and Rosalind Edwards

The Possibilities of Paradata case study concerns a project that explored the insights provided by the in-depth qualitative analysis of by-products of the research process—paradata. We analysed the marginalia written by field interviewers on archived copies of the paper questionnaires completed for the historically influential research, *Poverty in the United Kingdom 1967/68* (PinUK). Peter Townsend's survey played a pivotal and lasting role in the definition and measurement of poverty with international influence.

The PinUK research surveyed 3566 households, comprising data about 9584 household members in 634 areas. The notes recorded by the survey fieldworkers in the margins of the paper booklets used at the time comprised a body of written qualitative material that was beyond the capacities of our research team to read and analyse in depth as a complete set of data. So, our process involved selecting bundles of PinUK booklets from particular geographical areas: affluent, seaside, minority ethnic migration, and poor. Each of us reviewed our allocated bundles for booklets that contained large amounts of written marginalia and undertook a preliminary thematic overview as we went through them. By this process we selected 69 of the survey booklets overall for further analysis. Each of these selected booklets was then thematically analysed systematically by two of the team. This process identified seven main types of paradata around amplification, justification, and explanation of coding, evaluations and debriefings of the informants and the information they provided, and standpoints on social and political context. These procedures resonate with the aerial survey, geophysical survey, and test pit sampling approaches discussed in Chaps. 4, 5, and 6.

Our final element of our analytic process speaks to the deep excavation element of the breadth-and-depth method discussed in this chapter. We theoretically sampled six booklets by area, field interviewer and type of household, for in-depth narrative analysis. In two cases this analysis was by the team as a whole, and in the remaining four by pairs from the research team followed by discussion with all the team.

Our project took a constructionist perspective, viewing the PinUK fieldworkers and survey informants as making meaning within particular socio-historical contexts and thus as co-constructing the marginal paradata. The narrative method that we employed involved close reading of all the paradata in the survey booklet cases, with attention to the 'genre' of the 'story' constructed by the interviewer in their marginalia and how the interviewer positioned themselves narratively, as well as shifts in voice and tone. We also examined the dynamics of the narratives to see how they were built up and repeated, and inconsistencies and non-sequiturs. This in-depth narrative analysis provided insights into the interview process. It added important nuances to the thematic analysis in showing that the same theme could be addressed through different genres and positions. For example, some field interviewers positioned themselves as skeptical detectives uncovering the informants' circumstances, others treated informants' accounts as jigsaw puzzles to be pieced together to make sense. Other field interviewers positioned themselves as dispassionate observers, and some narratives functioned to position the PinUK interviewers as diligent fieldworkers. The tensions and labour of the field interviewers' work was revealed.

References

Elliott, H., Edwards, R., Phoenix, A., & Boddy J. (2015). *Narrative analysis of paradata from the Poverty in the UK survey: A worked example*. National Centre for Research Methods Working Paper 1/15. https://eprints.soton.ac.uk/375366/1/narrative_analysis_paradata.pdf

Phoenix, A., Boddy, J., Edwards, R., & Elliott H. (2017). "Another long and involved story": Narrative themes in the marginalia of the *Poverty in the UK* survey. In R. Edwards, J. Goodwin, H. O'Connor, & A. Phoenix (Eds.), *Working the paradata, marginalia and fieldnotes: The centrality of by-products of social research* (pp. 61–76). Edward Elgar.

*The project was funded as part of the ESRC National Centre for Research Methods programme, grant number ES/L008351/1.

Case Study 7.2 Growing Up on the Streets*

Lorraine van Blerk, Janine Hunter and Wayne Shand

Growing up on the streets is a big qual data set that documents the lives of children and youth living on the streets in diverse African cities. Data collection took place over a three-year period (2012 to 2016) in Accra, Ghana; Bukavu, Democratic Republic of Congo; and Harare, Zimbabwe (Growing up on the Streets, 2013, 2014).

The research was participatory and 18 street youth researchers, six in each city, engaged in ethnographic research among a network of 10 or more of their peers living in street settings (n=229). Applying a 'capability' approach (Sen, 1999; Nussbaum, 2000), adapted for young people living on the streets (van Blerk et al., 2016), ten capabilities were identified with street children and youth through a participatory process during the research design phase. Every week for three years the street youth researchers recorded ethnographic reports (ERs) in informal interviews with a streetworker/project manager who was known and trusted on the streets. The ten capabilities formed the framework for data collection and analysis; ERs were also free flowing and responsive to new daily challenges.

ERs were transcribed and translated into English locally and then logged, coded in NVivo, and cleaned at the University of Dundee, UK. The ten capabilities formed headline or 'parent' nodes (codes in NVivo); data is coded against clusters of 'child' nodes capturing all aspects of life growing up on the streets and individuals' case nodes; free nodes captured nuances such as gender-specific issues. In total data was coded over 130,000 times across 73 child and free nodes, plus case nodes, to facilitate longitudinal analysis of young people's lives; to explore the challenges they faced, the constrained choices they make, the resilience they demonstrate, and to provide nuanced understandings of the temporalities of their relationships, socio-spatial survival strategies, and future hopes.

A data set comprising 2478 fully anonymised and uncoded ERs is published on the UK Data Service ReShare (van Blerk et al., 2020); the largest in-depth repository on the lives of street living young people. Our initial longitudinal analysis took a narrative approach to exploring the undulations of life for young people on the streets. This approach simultaneously applied a case histories approach (Thomson, 2007) focusing on the six youth researchers in Accra. The scale of the data set meant that rather than applying a systematic selection of temporal points, the coded data could be dipped into cross-tabulating the capabilities framework applied with the before, during and after of major events in their individual lives as well as catastrophic events on the streets, including a fire in the market and two periods of severe flooding. This enabled an exploration of their resilience and the capabilities applied across the full range of capability themes. Ongoing analysis will follow this approach for all cities and young people. In addition, the data set lends itself to intersection analysis of (for example) individual capabilities and participants' age, gender or city; or to assess aspects of their lives in general (e.g. access to healthcare) or specifically (e.g. use of on-street medicines).

The size and scope of this big qual data set offers vast potential to analyse and understand both cross-cutting issues and specific nuances of growing up on the streets, as well as the ever-changing nature of street life alongside young people's continual presence in their cities.

References

Growing up on the Streets. (2013). *Briefing paper 1: Research principles*. StreetInvest/University of Dundee. https://doi.org/10.20933/100001136

Growing up on the Streets. (2014). *Briefing paper 2: Methodology*. StreetInvest/University of Dundee. https://doi.org/10.20933/100001137

Nussbaum, M. C. (2000). *Creating capabilities: The human development approach*. Harvard University Press.

Sen, A. (1999). *Development as freedom*. Oxford University Press.

Thomson, R. (2007). The qualitative longitudinal case history. *Social Policy & Society, 6*(4), 571–582.

van Blerk, L., Shand, W., & Shanahan, P (2016). Street children as researchers: Critical reflections on a participatory methodological process in the "growing up on the streets" research project in Africa. In L. Holt, R. Evans, & T. Skelton (Eds.), *Methodological approaches, geographies of children and young people* (Vol. 2, pp. 159–178). Springer.

van Blerk, L., Shand, W., Shanahan, P., & Hunter, J. (2020). *Growing up on the streets: Research with and for young people on the streets, 2012–2016*. Study number: 854123. Persistent identifier: https://doi.org/10.5255/UKDA-SN-854123

*Growing up on the streets acknowledges the generous funding received from Backstage Trust, which enabled the production of a big qual longitudinal data set on street life.

Case Study 7.3 Abductive Coding: The QUALIDEM Project

Claire Dupuy, Virginie Van Ingelgom and Luis Vila-Henninger.

In our research project QUALIDEM, we mobilize a related, yet original, approach to the Breadth-and-Depth Method. We abductively code existing material on citizens' perceptions of democracy and public policies from four qualitative data sets to develop abductive inference through in-depth interpretive analysis and, eventually, theory building (Vila-Henninger et al., 2022). In our abductive coding approach, researchers manually code multiple qualitative data sets that were collected by primary researchers and have been compiled into a single data set in order to answer research questions that span individual qualitative data sets. The new research questions theoretically drive data assemblage for secondary qualitative analysis—thus drawing on the opportunities created by comparison (Halford & Savage, 2017).

Methodologically, our approach combines traditional approaches to abduction (Pierce, 1934) with qualitative approaches (Timmermans & Tavory, 2012; Tavory & Timmermans, 2014). In particular, the qualitative branch of abduction builds theory by engaging with the data via qualitative coding and identifying theoretically anomalous cases. Timmermans and Tavory (Timmermans & Tavory, 2012; Tavory & Timmermans 2014) highlight the methodological steps that are conducive to the implementation of the abductive logic of data analysis. These three steps consist of "Revisiting the phenomenon", "Defamiliarization", and "Alternative casing". In the context of our research project, we suggest that each step raises specific issues and is performed through specific operations (Vila-Henninger et al., 2022).

Concretely, building on these three steps, our approach develops three phases for abductive coding: (1) Generating an Abductive Codebook, (2) Abductive Data Reduction, and (3) Abductive Qualitative Analysis. We start with an initial deductive, theoretically-driven codebook, which then allows us to develop inductive codes for cases in the data that are "theoretically anomalous"—or not present in the initial theoretically-driven deductive codebook. By this, we mean that human coders start with a deductive codebook and then through the process of coding, build the codebook—and by extension build theory—by developing data-driven inductive codes that document theoretically anomalous cases. The approach then proceeds through data reduction with Qualitative Data Analysis Software (QDAS) to combine codes in order to operationalize phenomena that span data sets (Evers et al., 2020). This stage of the analysis allows for further inductive coding and facilitates a final round of qualitative analysis that researchers perform manually. To summarize, our approach builds on and expands the Breadth-and-Depth Method by mobilizing abductive coding and inference.

References

Halford, H., & Savage, M. (2017). Speaking sociologically with big data: Symphonic social science and the future for big data research. *Sociology, 51*(6), 1132–1148.

Evers J, Caprioli, M U, Nöst, S., & Wiedemann, G. (2020). What is the REFI-QDA standard: Experimenting With the transfer of analyzed research projects between QDA software. *FQS*, *21*(2), Art. 22.
Peirce, C. (1934). *Collected papers of charles sanders peirce. Vol. 5, pragmatism and pragmaticism* (C. Hartshorne, & P. Weiss, Eds.). Harvard University Press.
Tavory, I., & Timmermans, S. (2014). *Abductive analysis: Theorizing qualitative research*. University of Chicago Press.
Timmermans, S., & Tavory, I. (2012). Theory construction in qualitative research from grounded theory to abductive analysis. *Sociological Theory*, *30*(3), 167–186.
Vila-Henninger, L., Dupuy, C., Van Ingelgom, V., Caprioli, M. U., Teuber, F., Pennetreau, D., Bussi, M., & Le Gall., C. (2022). Abductive coding: Theory building and qualitative (re)Analysis. *Sociological Methods and Research*. https://journals.sagepub.com/doi/full/10.1177/00491241211067508

*This research was funded by the project ERC Starting Grant Qualidem—Eroding Democracies. A qualitative (re-) appraisal of how policies shape democratic linkages in Western Democracies. The Qualidem project is supported by the European Research Council (ERC) under the European Union's Horizon 2020 research and innovation programme (grant agreement 716,208).

References

Andrews, M., Squire, C., & Tamboukou, M. (2013). *Doing narrative research* (2nd ed.). Sage.
Bartlett, R. (2012). Modifying the diary interview method to research the lives of people with dementia. *Qualitative Health Research, 22*(12), 1717–1726.
Braun, V., & Clarke, V. (2006). Using thematic analysis in psychology. *Qualitative Research in Psychology, 3*(2), 77–101.
Braun, V., & Clarke, V. (2020). Can I use TA? Should I use TA? Should I not use TA? Comparing reflexive thematic analysis and other pattern-based qualitative analytic approaches. *Counselling and Psychotherapy Research*. https://doi.org/10.1002/capr.12360
Campbell, R., Britten, N., Pound, P., Donovan, J., Morgan, M., Pill, R., & Pope, C. (2006). Using meta-ethnography to synthesise qualitative research. In J. Popay (Ed.), *Moving beyond effectiveness in evidence synthesis: Methodological issues in the synthesis of diverse sources of evidence* (pp. 75–82). National Institute for Health and Clinical Excellence.
Charmaz, K. (Ed.). (2014). *Constructing grounded theory* (2nd ed.). Sage.
Coffey, A., & Atkinson, P. (1996). *Making sense of qualitative data: Complementary research strategies*. Sage.
Drewett, P. (1999). *Field archaeology: An introduction*. Routledge. https://bit.ly/3mBLZ9S
Edwards, R., & Weller, S. (2012). Shifting analytic ontology: Using I-poems in qualitative longitudinal research. *Qualitative Research, 12*(2), 202–217.
Edwards, R., Weller, S., Davidson, E., & Jamieson, L. (2021). Small stories of home moves: A gendered and generational breadth-and-depth investigation. *Sociological Research Online*. https://journals.sagepub.com/doi/full/10.1177/13607804211042033
Elliott, J. (2008). The narrative potential of the British birth cohort studies. *Qualitative Research, 8*(3), 411–421.
Emmel, N. (2013). *Sampling and choosing cases in qualitative research: A realist approach*. Sage.
Fairclough, N. (2010). *Critical discourse analysis: The critical study of language*. Longman.
Flick, U. (2018). *Doing grounded theory*. Sage.

Fontaine, C. M., Baker, A. C., Zaghloul, T. H., & Carlson, M. (2020). Clinical data mining with the listening guide: An approach to narrative big qual. *International Journal of Qualitative Methods, 19*, 1–13.
Fritz, R. L., & Vandermause, R. (2017). Data collection via in-depth email interviewing: Lessons from the field. *Qualitative Health Research, 28*(10), 1640–1649.
Giddens, A. (1992). *The transformation of intimacy: Sexuality, love, and eroticism in modern societies*. Polity Press.
Gilligan, C., Spencer, R., Weinberg, M. K., & Bertsch, T. (2003). On the listening guide: A voice-centred relational method. In P. M. Camic, J. E. Rhodes, & L. Yardley (Eds.), *Qualitative research in psychology: Expanding perspectives in methodology and design* (pp. 157–173). American Psychological Association.
Grbich, C. (2007). *Qualitative analysis: An introduction*. Sage.
Hookway, N. (2008). 'Entering the blogosphere': Some strategies for using blogs in social research. *Qualitative Research, 8*(1), 91–113.
Liddicoat, A. J. (2011). *An introduction to conversation analysis* (2nd ed.). Bloomsbury.
Mason, J. (2002). *Qualitative researching* (2nd ed.). Sage.
Mauthner, N., & Doucet, A. (1998). Reflections on a voice-centred relational method: Analysing maternal and domestic voices. In J. Ribbens & R. Edwards (Eds.), *Feminist dilemmas in qualitative research: Private lives and public knowledge* (pp. 119–146). Sage.
Neale, B. (2016). *What is qualitative longitudinal research?* Bloomsbury Academic.
Neufeld, L. M., Andrade, E. B., Ballonoff Suleiman, A., Barker, M., Beal, T., Blum, L. S., Demmler, K. M., Dogra, S., Hardy-Johnson, P., Lahiri, A., Larson, N., Roberto, C. A., Rodríguez-Ramírez, S., Sethi, V., Shamah-Levy, T., Strömmer, S., Tumilowicz, A., Weller, S., & Zou, Z. (2022). Food choice in transition: Adolescent autonomy, agency, and the food environment. *The Lancet, 399*, 185–197.
Reissman, C. K. (1993). *Narrative analysis, qualitative research methods series 30*. Sage.
Ritchie, J., & Lewis, J. (Eds.). (2003). *Qualitative research practice: A guide for social science students and researchers*. Sage.
Rosenthal, G. (1993). Reconstruction of life stories: Principles of selection in generating stories for narrative biographical interviews. In R. Josselson & A. Lieblich (Eds.), *The narrative study of lives* (Vol. 1). Sage.
Rosenthal, G. (2018). *Interpretive social research: An introduction*. Göttingen University Press. https://library.oapen.org/bitstream/handle/20.500.12657/27538/1002466.pdf?sequence=1&isAllowed=y
Sandelowski, M., & Barroso, J. (2006). *Handbook for synthesising qualitative research*. Sage.
Sidnell, J. (2010). *Conversation analysis: An introduction*. Wiley.
Taylor, S. (2013). *What is discourse analysis?* Bloomsbury Academic.
Yanow, D., & Schartz-Shea, P. (2006). *Interpretation and method: Empirical research methods and the interpretive turn*. M.E. Sharp.

Part IV

Reflecting on the Implications of Large-Scale Qualitative Analysis

Ethics and Practice

8

8.1 Introduction

The preceding chapters outlined the rationale for and steps involved in conducting large-scale qualitative analysis using the breadth-and-depth method. In thinking about the ethical issues concerned with big qual analysis and how we might conceive of good practice, it is essential to look beyond the ethical procedures and formal processes that our universities or research institutions may have in place. Rather, we need to focus on the situated nature of ethical dilemmas, the context in which issues arise, and how they evolve over time (Pascoe Leahy, 2021; Lyle et al., 2022). The aim of this chapter is to provide a consolidated overview of the key challenges and potential mitigations *across* the process. Some of these issues are not unique to big qual or to the use of the breadth-and-depth method but are likely to be encountered when employing other approaches to qualitative secondary analysis (QSA). That being said, big qual analysis can exacerbate some of these concerns.

We argue that any form of big qual analysis must be mindful of two sets of ethical issues on which we elaborate in this chapter. In brief, the first set relates to broader concerns regarding the epistemic privileging of quantity inherent to big data debates (Mills, 2018) and the use of technology to overcome the challenges of handling data at volume. 'More' or 'bigger' is not necessarily better, and the tools used generate new ethical challenges for qualitative researchers. The second set focuses on the fair and equitable treatment of qualitative data and the investments of all those involved, from research participants and researchers, through to archivists and data managers, when working at scale, including when using the breadth-and-depth method.

In this chapter, we will:

- provide an overview of the key ethical issues concerning large-scale qualitative analysis
- highlight the ways in which such issues align with or differ from concerns commonplace in qualitative work

© The Author(s), under exclusive license to Springer Nature Switzerland AG 2023

S. Weller et al., *Big Qual*, https://doi.org/10.1007/978-3-031-36324-5_8

- connect to wider debates regarding the sharing and reuse of qualitative data
- offer examples of ethical dilemmas faced in bringing together multiple qualitative data sets
- discuss qualitative integrity
- consider what might constitute good secondary analytic practice and offer examples of a range of approaches

8.2 Ethical Practice and Constructing a Corpus

8.2.1 Issues of Data Sovereignty

Data sovereignty is often used to refer to compliance with laws in the country in which data was generated. In pooling data sets from different sources, it is likely that you will need to grapple with regulation around data ownership and data protection legislation across multiple national contexts. Even with the appropriate data sharing agreements in place, a key issue of relevance to big data debates and QSA is access. This concerns *who* has access to large volumes of data, along with the resources necessary to analyse it (Mills, 2018). With any approach to secondary analysis, both qualitative and quantitative, comes the danger that, if not practised with care, the contributions of researchers and/or participants could, albeit unintentionally, be exploited. For instance, studies that involve combining data generated in different national contexts must be alert to the possibility that the work of the original researchers could be reduced to 'data generators' if opportunities to collaborate on any reuse endeavours are not evident. In Chap. 6, Case Study 6.2—"Young people and food choice in transition"—discussed the use of the breadth-and-depth method to analyse 11 data sets, some of which were the outcome of the international, interdisciplinary "Transforming Adolescent Lives Through Nutrition" (TALENT) initiative. The programme was founded on shared learning, mutual visits, and collaborative analysis (Hardy-Johnson et al., 2021). Even with this approach, the analysis was led by researchers in the UK who had access to the computational tools necessary for steps two and three.

Data sovereignty may also be considered in relation to self-governance. Of significance is Indigenous data sovereignty, which has "become an increasingly relevant topic as big data, open data, open science, and data reuse become an integral part of research and institutional practices" (Carroll et al., 2020, p. 3). Data sovereignty in this respect refers to participant control over their own stories rather than institutional bodies, more collective notions of ownership (Carroll et al., 2020), and benefit-sharing with respect to secondary analysis (Keikelame & Swartz, 2019; Carroll et al., 2020). Of particular interest is "CARE Principles for Indigenous Data Governance" (Collective Benefit, Authority to Control, Responsibility, and Ethics) (Research Data Alliance International Indigenous Data Sovereignty Interest Group, 2019). In developing approaches to big qual analysis, we must aim to develop practices which hold central notions of collective benefit and respectful research relationships, and ensure data control is not reduced to regulation (Carroll et al., 2020).

Indigenous perspectives on data sovereignty take us further, however, towards the decolonisation of research, in raising issues about ethics and practice that transform fundamentally the whole nature of research, in terms of what counts as knowledge and who produces, owns, uses, and benefits from it. Embarking upon decolonisation calls for researchers to make explicit the epistemological, ontological, and methodological standpoints that inform their research processes. It involves researchers engaging in critical reflexivity regarding their assumptions and interpretations, exercising reciprocity and respect for self-determination of research participants, embracing 'Other(ed)' ways of knowing, and embodying a transformative practice (Thambinathan & Kinsella, 2021). As we note at various points in this book, researchers utilising the breadth-and-depth method for working with large amounts of qualitative data need to be reflexive about their logic of enquiry, theoretical and methodological standpoints, and how they interlink. Decolonising and Indigenous approaches to research raise sets of questions around who owns the research issues, who initiates them, in whose interests the research is carried out, who has control of research, how power relations and decisions are negotiated in creating knowledge, who the research is for, what counts as knowledge, who is transformed by it, and whose is the authorial voice. Indigenous scholars have pointed to the narrow focus on Indigenous and marginalised peoples as lacking and in need of improvement in mainstream epistemologies and methodologies, especially prevalent in pre-formed secondary data sets, rather than on those who benefit from their impoverishment and how they sustain their power and maintain privilege (e.g. Walter & Andersen, 2016).

There are debates about the extent to which non-Indigenous researchers are able to work with Indigenous perspectives (e.g. Chilisa et al., 2017; Tuhiwai Smith, 2021), mainly in relation to the process for primary research. Nonetheless, there are thought-provoking issues raised for researchers working with secondary qualitative data, whatever their affiliation and background. Drawing on an African-based relational ontology, for example, Chilisa et al. (2017) present an Indigenous-centred epistemological and ethical framework that they argue both Indigenous and non-Indigenous researchers can adapt to guide research. As they point out, "a relational ethical framework invites researchers to see 'self' as a reflection of the researched 'Other', to honour and respect the researched as one would wish for oneself, and to feel a belongingness to the researched community without feeling threatened or diminished" (p. 328). At the core of their approach is a series of the sorts of questions we have outlined above, that ask researchers to reflect on the extent of various aspects of the social relevance and transformative intent of their research purposes, as well as the systems of understanding and worldviews held by the research participants and the power differential with mainstream academic frameworks. Such critical issues may apply as much to each of the steps of the breadth-and-depth approach to working with large amounts of qualitative secondary data, and the method as a whole, as they do to the generation and analytic purposes of primary material, whoever the research participants.

8.2.2 Overcoming Exclusion in the Archives

Whilst there are many benefits to using data made available through centralised repositories, we need to be attentive to how institutionalised approaches to archiving govern not only the nature of the data archived, but to what researchers have access (Sherren et al., 2017; Glenna et al., 2019). If the voices of seldom listened to individuals and communities are under-represented or absent from the material that is curated, then reusing such material will only perpetuate these inequalities. This is a concern that stretches across each step of the breadth-and-depth method. In Chap. 4 we described the process of systematically auditing available data sets to determine whether they are of an appropriate nature, quality, and 'fit' with the research topic. An essential part of this process may involve determining the diversity of the data (sets), including individuals, collectives, and/or communities not represented, that will then constitute the corpus. It may be that the relevant metadata is not available, either because it was not recorded or because it is of a sensitive nature and, if used, could permit re-identification. There is an inherent danger that if the breadth-and-depth method is employed only with data from institutional repositories, then any future analysis will just reproduce inequalities. Indeed, we had some experience of this in our own work combining data sets housed in the Timescapes Data Archive. Whilst we had desired a culturally and ethnically diverse sample, some of the original studies were conducted in locales with relatively homogeneous populations with the intended samples stratified along different lines. In some instances, metadata regarding aspects of participants' identities were missing or incomplete.

Breadth-and-depth analysts can help to rectify this through the careful selection of data sets to construct a corpus, by including material currently not housed in archives, through the sample size and logic employed (with specific attention to who and how many participants are included), via the selection of cases for deep excavation, and by depositing new assemblages for others to use. The breadth-and-depth method can also be used as a precursor to further empirical work and to help identify under-researched voices.

A further challenge associated with sourcing archived data for breadth-and-depth analysis is that, whilst there are an increasing number of repositories, large-scale infrastructure is not equally distributed globally (Corti, 2007). A reliance on archived material could exacerbate the dominance of research from European and North American contexts and perspectives to the exclusion of settings where the necessary digital infrastructure does not exist. More centralised approaches to archiving are but one model, and a solution may be to look beyond larger-scale repositories to co-constructed community archives (Moore et al., 2017; Thomson & Berriman, 2021), or to the creation of networks and data sharing agreements with others. Case Study 8.2 provides an interesting example. That being said, authors such as Glenna et al. (2019) have argued that "… big-qualitative data could be used to extend research and analysis opportunities to communities that do not have adequate access to the resources and funding needed for the collection of primary data" (p. 576). An advantage of the breadth-and-depth method is the possibility of diversifying data sets by developing a carefully constructed corpus, but this needs to be carried out with care. The "Young people and food choice in transition" analysis, detailed throughout this volume, is a prime example of this in that existing material from multiple projects and international contexts was brought together via those involved in a set of connected projects and networks.

8.3 Working with Integrity at Scale

8.3.1 Research Integrity

Throughout the preceding chapters we have discussed the importance of conducting big qual analysis with integrity, in an honest, responsible, and transparent manner that inspires trust in the approach and outcomes. Methodological and interpretative integrity are key to qualitative work. Methodological integrity is more than about selecting the most appropriate methods to answer a research question or set of questions. Rather it involves having an approach that demonstrates coherence between ontology, epistemology, and how interpretations are derived (Neale, 2021, p. 350). Furthermore, interpretative integrity is concerned with researchers' interpretations of qualitative data and whether such work is conducted with "veracity, verisimilitude, trustworthiness and honesty" (Neale, 2021, p. 350). Integrity in qualitative research cannot be measured or judged by compliance with guidelines but is "contingent upon context and situation rather than abstract principles" (Watts, 2008, online).

Guided by their philosophical position, researchers will, in practice, take different approaches to ethical decision-making (Tisdale, 2004). Integrity is relational between researchers and participants, original and new researchers/teams, and researchers and research institutions (Watts, 2008). Reusers of qualitative data trust that the original researchers worked with methodological and interpretative integrity and have conducted studies responsibly and with honesty (Roulston, 2017; Neale, 2021). Similarly, researchers, teams, and participants involved in the generation of data trust that any reusers handle data responsibly and with care and respect the efforts of those who co-produced/made the data available for reuse. Common practices will shift over time and may differ between contexts.

For the breadth-and-depth method, this means that analysts ought to explore, as far as is feasible, the philosophical underpinnings and ethical approaches of the original teams. It also means being reflexive and open about your own assumptions and positioning towards the data. Integrity concerns looking beyond the use-value of existing data and treating with care and respect the investments made by the original researchers and participants. In so doing, we need to transcend ideas of researchers as stewards tasked with managing data, to think of ourselves as custodians tasked with caring for and about data in a manner that is mindful of the voices of those within the data as well as those who may be missing or seldom heard. Working with integrity means acknowledging and, where possible, mitigating against the shortcomings of the computational tools employed in step two, recognising how the assumptions underlying approaches to text mining that rely on algorithm-based tools can shape outcomes. Moving iteratively between the abstracted breadth analysis and the minutiae of the data to reconnecting with context is key to achieving this. It also concerns interpreting and producing new knowledge from existing data sets with integrity.

8.3.2 Establishing the Nature of Consent

One of the most prominent ethical concerns in the reuse of any data, including qualitative material, is consent. Whether we use data sets housed in local archives or national repositories, or whether we combine material from multiple projects, issues of consent and the permissions afforded by research participants are likely to shape the material selected to create a new data assemblage (step one). Contemporary data that has been archived will probably have been deposited following a juridical model of signing consent forms and in compliance with the relevant data protection policies and laws. Drawing on biomedical research, notions of broad or enduring consent, where future uses that are different from the original research objectives are either not specified or unknown, are commonplace (Zarate et al., 2016; Neale, 2021). For data that sits outside of an archive or was generated before the advent of formalised consent processes, understanding to what participants may have agreed is less clear cut. In these instances, the ways in which potential future uses were articulated (or not), how participants conceive of data storage, the immortality of some types of data (Lyle et al., under review), and re-purposing of data will be less clear or unknown. Furthermore, data may become more or less sensitive over time (Glenna et al., 2019). Consent forms are often prescriptive in nature and offer little insight into the negotiations that may have taken place between researchers and participants. Furthermore, the nature of potential future uses is challenging and shifts over time. The breadth-and-depth method, for example, is only feasible due to developments in digital infrastructures and computational tools, and the pooling of data sets is a possibility that participants may never have envisaged 20 years ago. For users of existing data, then, it is important to be cognisant of such issues and uncertainties, and to be diligent in terms of checking data sources and origins. The challenges of establishing consent are described well in Case 4.4. Purcell and Maxwell spent time during the funding application to ensure consent was in place for reuse across the 11 projects. However, the practicalities of agreeing and signing off institutional data sharing agreements, and facilitating secure transfer of data sets, were time-consuming.

8.3.3 Risks of Data Linkage

In bringing together multiple data sets it is, of course, possible that any original assurances of confidentiality through, for example, the removal of direct and indirect identifiers may be breached, albeit inadvertently. Although data held in repositories is likely to have been anonymised, it is feasible that by connecting and bringing data sets into conversation the identities of people, places, institutions, or organisations may be more identifiable. This may be the case when linking research data or when bringing together different aspects of metadata (Hughes & Tarrant, 2020). These concerns link to broader debates within, for example, health literature. Glenna et al. (2019), for instance, highlight the risks of participants living with rare conditions being identified in genomic research. Equally, this is a possibility for those living in distinct communities, circumstances, or organisations. As Zarate et al. (2016) suggest, the risk of re-identification can have implications for public trust in research. This is a contingency that needs to be considered across all steps of the breadth-and-depth method from making decisions about what to include and the bringing together of metadata (step one) through to in-depth analysis (step four). We also need to think about the risks of re-identification through the, perhaps inadvertent, linking of research data to other traces of material beyond the assemblage, for example, on social media. Case Study 8.3 illustrates such challenges.

8.3.4 Centrality of Context

One of the main concerns in the pooling of qualitative data generated in different projects or programmes is context. In Chap. 4, we highlighted the simplistic distinction between primary and secondary analysis. Earlier debates about the feasibility and ethics of QSA were oriented around context, and whether researchers could analyse meaningfully data they had not played a part in generating (Mason, 2007). A key element of this argument concerns how qualitative data is conceived and the view of such material as co-constructured through the encounters and interactions between researchers and participants. In these terms, then, the data is not positioned as belonging to research participant(s) only; researchers' "investments and what they include of themselves, personally and professionally, also are bound up in the data" (Weller, 2022).

An issue with big data is analytic distance and the dangers of decontextualisation and misunderstanding the context in which it was generated (Harford, 2014). With large-scale qualitative analysis it can prove challenging both to access a sense of the contexts in which individual cases or data sets were generated, and to hold multiple contexts central when grappling with large volumes of data. A strength of the breadth-and-depth method is the iterative nature of the process and the movement between the broader surveying techniques (step two) and the in-depth elements (steps three into four). Many of the tools outlined in Chaps. 5 and 6 enable aspects of the context to remain visible during steps two and three. For instance, in the "Young people and food choice in transition" case study (Case Study 6.2), which brought together 11 data sets, the text mining software Leximancer (v4·51) was employed. Figure 8.1 illustrates the dashboard, with the output—a two-dimensional visual map—taking a central position, whilst on the right it is possible to view data extracts in the context of the wider document in which they feature. A summary of the case metadata is also visible, supporting researchers in remaining connected to the original context in which the data was generated. Nevertheless, it is essential

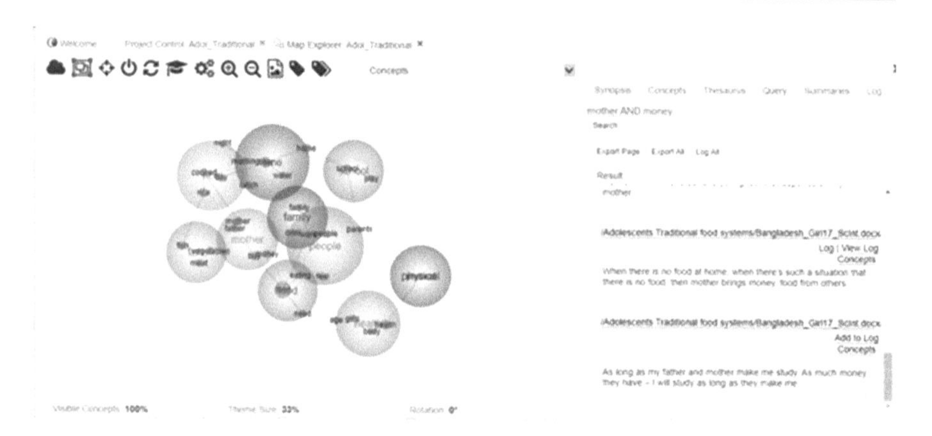

Fig. 8.1 Viewing extracts-in-context on the Leximancer dashboard

that context is not reduced to a feature of a particular software platform. Rather, the over-riding challenge of large-scale qualitative analysis is reconciling multiple contexts within the interpretative work. For instance, in the test pit sampling of step three this might involve systematically organising extracts by context within a particular theme, or in the deep excavations of step four this might entail working with whole cases, conducting multiple readings, and identifying and describing relationships and/or patterns of meaning.

8.3.5 Reliance on Computational Tools and Algorithms

In step two of the breadth-and-depth method, text-mining and semantic techniques are employed to help manage large volumes of data and to identify features of interest within the data landscape to excavate further. For us, using computational tools to conduct such analysis represented a departure from the approaches we had adopted previously and made us acutely aware of the ethical challenges in so doing. We concur with Glenna et al. (2019) that "Although new data gathering, storage, sharing, and analyzing technologies enable innovative approaches to qualitative research, they also create unique challenges for the responsible use of that data" (p. 561).

A key ethical issue for step two relates to the potential reliance on machine learning (employing data and algorithms to emulate human learning) and, often patented, algorithm-based tools to conduct semantic analysis. These algorithms are used to sift through thousands of iterations to see how words 'travel together' through a text or set of documents in order to pinpoint words that are co-located beyond what is considered random statistically. So much of what qualitative researchers strive for rests on taking a reflexive approach that both acknowledges and allows for differing ontological positions. However, the assumptions on which patented algorithms are founded are not transparent to the user. The danger, then, is that researchers may not be aware of or able to determine the basis of the worldviews on which they are constructed (Glenna et al., 2019). In discussing wider debates in, for instance, critical data studies, authors such as Glenna et al. (2019) contend that a reliance on algorithm-based tools can mask or perpetuate inequalities. The concern, then, is the role such tools play in mediating researchers' decision-making (Mittelstadt et al., 2016; Glenna et al., 2019). As part of steps two and three, it is essential that researchers reflect their own decision-making around, for example, inclusion or exclusion criteria for terms, the rationale for Boolean searches, the length of extracts selected, and the movement between extracts and extracts-in-context.

We were all too aware of these challenges in developing the breadth-and-depth method, particularly in terms of our experimental work with different tools during step two (see Chap. 5). As Mills (2018) argues, "It has been long recognized that despite their superior speed of mathematical computations, computers are unable to capture the subtlety, creativity, and personality that real human beings demonstrate across social contexts" (p. 600). For us, computational tools must not be employed uncritically, and text mining should not be regarded as the destination but rather as a staging post on the journey. Within the breadth-and-depth method, we regard the computational tools employed in step two as a means of sifting through and gleaning an overview of the assemblage, which will then aid more in-depth analysis. This process does not replace the interpretative work of the researcher, but rather enables new or different questions to be asked of a corpus of data. As such, this kind of work sits alongside, but not in competition with, smaller-scale work.

8.4 Working with Care

8.4.1 Caring for and About Researchers' Investments

Whilst there is a growing literature on ethical issues associated with QSA, the implications for researchers and research teams have received less attention. The policies that govern data sharing have been developed largely in response to quantitative data management strategies, which can make navigating the formalised, regulated, and institutionalised landscape of data sharing challenging for qualitative researchers (Weller, 2022). Such measures focus on procedural matters at the expense of considering what researchers invest in terms of time and energy, emotion, and intellectual creativity (Mauthner, 2012; Weller, 2022). For primary researchers, there is much work involved in data preparation and curation. In teams, this often falls to those in more junior positions who, in the context of wider reforms in higher education (HE), are increasingly likely to find themselves in precarious employment situations. Preparing data for reuse by others is often an under-valued task that may be completed at the expense of other career-enhancing endeavours (Weller, 2022). Despite the view of data as co-constructed, researchers are not afforded the same privileges as participants relating to anonymity, confidentiality, and so on. As we have argued elsewhere, "The identities of researchers and what they reflexively reveal of themselves, how they interact with participants, their techniques and approaches and the messiness of qualitative work are laid bare within the artefacts of qualitative data" (Weller, 2022, p. 1030). As such, with greater scrutiny comes the danger that researchers feel exposed personally or professionally, or that their investments are exploited, or reputation damaged (Gillies & Edwards, 2012).

If feasible/appropriate we encourage you to learn as much about the original researchers as possible and to study descriptive metanarratives, researchers' biographies, and project documents. If desirable and/or feasible, consider working with the original researchers, although we recognise that this might not be viable because the original researchers may not have the time or interest, or they may be untraceable or have passed away. Secondary users of the data may also be lacking in time or resources. Neale (2021) proposes three strategies for collaborative work: i) intermittent sharing via, for example, workshops; ii) affiliated sharing whilst projects are ongoing; and iii) partnership sharing through data pooling. Our own plans and final approach are outlined in Case Study 8.1. We also acknowledge that some may regard the engagement of primary researchers as an interference, instead viewing the data as embodying new knowledge or alternative insights, which do not require the explicit involvement of the original researchers. It may also be that the secondary analysis challenges the original researcher's interpretation. A classic example of this is Walkerdine and Lucey's (1989) re-analysis of data from Tizard and Hughes' (1984) study of young children learning at school and at home, which revealed the class-based assumptions about child-rearing embedded in the original analysis.

8.4.2 Shifting Connectedness to Data

Questions around data ownership often occur in discussions about QSA. For example, those preparing qualitative data for deposit in an archive are compelled to consider the legal and regulatory aspects. From this perspective, data ownership is conceived in terms of matters such as Intellectual Property (IP) and the ownership of data as residing with the institution where a researcher was employed at the time of research data creation, rather than with an individual researcher or team (Weller, 2022). This masks the complexity of qualitative data. A sense of ownership has meaning beyond regulatory structures and can emerge from the emotional and intellectual connections fostered through the temporal resources, commitments, and care researchers invest in the relationships that are central to the production of qualitative data (Weller, 2022; Weller & Edwards, 2022). A sense of ownership is not just confined to the empirical work in which a researcher has been directly engaged. As Case Study 8.4 illustrates, those undertaking QSA also often feel a growing sense of connection to the material with which they are working. The iterative nature of the breadth-and-depth method and, in particular, the emphasis in steps three and four, of starting to explore the material in more depth and then the process of undertaking deep excavations, is where a growing sense of ownership is likely to emerge.

8.4.3 Researcher Well-Being

Discussions regarding the analysis of big data rarely include attention to researcher well-being. An issue that is pertinent across all steps is the impact of engaging with sensitive or traumatic data on the researcher, especially when reading accounts where emotional or emotive issues were not anticipated. Ballard (2020) also talks of the distress she experienced in working with diary data that contained raw and candid accounts from participants awaiting the outcome of genetic testing for Huntingdon's disease. This was exacerbated during the pandemic as she was conducting the analysis alone, at home and without the usual support network of colleagues. Ballard couches her reflections in terms of the vicarious trauma experienced through in-depth and multiple readings of challenging accounts. As she notes, these kinds of ethical issues rarely feature in institutional processes for ethical approval. The impact of engaging with such accounts is an issue of relevance to all QSA. It is, therefore, important that time to reflect on the effects of the material with which you have engaged is built into studies, and that researchers are supported/have opportunities to engage in networks and practices of support.

8.4.4 Caring about Environmental Impact

A broader ethical implication of the growth in big data, the open science agenda, and digital preservation relates to sustainability (Lucivero, 2020; Samuel & Lucivero, 2020). A more pragmatic advantage of reusing existing digitised qualitative data is the reduction in travel associated with in-person empirical research, although the global pandemic has undoubtedly altered research practices (Nind et al., 2021). Whilst less travel might reduce a researcher's carbon footprint, the environmental costs of creating and sustaining the infrastructure and resources necessary for long-term data preservation and for processing data are often overlooked. Such costs include the consumption of non-renewable energy, raw material usage, e-waste disposal, and the production of carbon dioxide associated with the construction and maintenance of data storage facilities and data processing (Lucivero, 2020, Samuel & Lucivero, 2020). Although qualitative data sets of the nature outlined in Chap. 3 represent but a small part of this, it is important to understand that storing and processing data have environmental impacts. Again, in discussing big data more broadly, Mills (2018) urges further caution about contributing to "a data-overloaded world" (p. 596). The environmental impact of our own research endeavours warrants due consideration, and in using the breadth-and-depth method this is likely to feature in decision-making around developing a rationale for the research at the outset and, during step one, when sourcing data. For instance, breadth-and-depth researchers may wish to consider the environmental credentials and policies of the archives from which they wish to source data (e.g. https://kdl.kcl.ac.uk/our-work/archiving-sustainability/).

8.4.5 Ethic of Care Framework

Finally, in thinking across the breadth-and-depth process we have, in our own practice, found it useful to draw on the ethic of care literature and to think about how we might infuse 'habits of care' into processes of data sharing (Weller, 2022). This can be extended to the pooling of multiple qualitative data sets and the use of the breadth-and-depth method. As we have outlined in more detail elsewhere (Weller, 2022), this might include caring for and about investments in qualitative data production by recognising the emotional and temporal investments made by the original researchers/teams in bringing data sets to fruition, including the labour involved in curating materials for reuse. It could also entail thinking with care about researchers' differing positions in the wider HE landscape by understanding that the task of archiving often falls to more junior researchers, who are more likely to inhabit precarious positions in the HE labour market, and may complete archiving tasks at the expense of other career-enhancing activities. It also concerns prioritising the place of researchers in the production of qualitative data through working respectfully with secondary data in a manner that acknowledges the presence of the researchers' identities, professional reputations, and personal lives as an integral part of the data within that material. A relational view of care and responsibility in research teams is also necessary to allow for more collegial and equitable practices of data sharing. Finally, it is vital that we think carefully about where and how we source qualitative data, looking beyond large-scale repositories.

8.5 Resources

The following papers provide some useful ethical reflections of relevance:

Ballard, L. (2020). *Wellbeing of psychologists during Covid-19: Impact on research* https://www.youtube.com/watch?v=u8DGEW16GZ4

Carroll, S. R., Garba, I., Figueroa-Rodríguez, O. L., Holbrook, J., Lovett, R., Materechera, S., Parsons, M., Raseroka, K., Rodriguez-Lonebear, D., Rowe, R., Sara, R., Walker, J. D., Anderson, J., & Hudson, M. (2020). The CARE Principles for Indigenous Data Governance. *Data Science Journal*, *19*(1), 43. https://doi.org/10.5334/dsj-2020-043

Corti, L., Van den Eynden, V., Bishop, L., & Woollard, M. (2020). *Managing and sharing research data: A guide to good practice*. Sage.

European Commission. (2016). *Guidelines on FAIR data management in horizon H2020*. https://ec.europa.eu/research/participants/data/ref/h2020/grants_manual/hi/oa_pilot/h2020-hi-oa-data-mgt_en.pdf

Edwards, R., & Weller, S. (2015). Ethical dilemmas around anonymity and confidentiality in longitudinal research data sharing: The case of Dan, In M. Tolich (Ed.), *Qualitative ethics in practice*. Left Coast.

Morrow, V. (2017). The ethics of secondary data analysis. *Big Qual Analysis Resource Hub*. http://bigqlr.ncrm.ac.uk/2017/03/29/guest-blog-9/

TRUST—Protecting San indigenous knowledge—From a research contract to a San code of ethics. https://youtu.be/HOdw3mv7JSo
Weller, S. (2019). *Collaborating with original research teams: Some reflections on good secondary analytic practice.* http://bigqlr.ncrm.ac.uk/2019/03/06/post26-dr-susie-weller-collaborating-with-original-research-teams-some-reflections-on-good-secondary-analytic-practice/
Weller, S., Davidson, E., Edwards, R., & Jamieson, L. (2019). *Big qual analysis: Teaching data set.* University of Leeds, UK Timescapes Archive. https://doi.org/10.23635/14
Weller, S. (2022). Fostering habits of care: Reframing qualitative data sharing policies and practices. *Qualitative Research.* https://doi.org/10.1177/14687941211061054

▶ Case Study 8.1 Collaborating with the Original Research Teams

Susie Weller

In developing the 'Working across qualitative longitudinal studies: A feasibility study looking at care and intimacy' project (https://bigqlr.ncrm.ac.uk/), we were mindful of all that our Timescapes colleagues had invested in their projects and in the long-term relationships they had developed with participants. Even though the original teams had archived their data to enable re-use, guided by an ethic of care, we felt we ought to consult them about our plans and invite them to be involved if they wished. We had worked closely with the original teams previously and we had contributed to developing the processes for curating the data for re-use. Accordingly, we contacted colleagues to inform them of the purpose of our new study and our plans to use the Timescapes data. In the early stages, we liaised with individuals via email, asking project-specific questions about, for instance, the research context, data set structure and their own analysis. Whilst our intention was to be inclusive, in practice we liaised with only one or two members of the original team; those with whom we had strong professional relationships.

Later, we took a more formalised approach inviting former colleagues to complete an online consultation with questions asking them about their changing connection to the data, feelings and concerns about data sharing and reuse, and the forms of consultation or connection (if any) they would consider appropriate or valuable. We soon came to realise that, whilst our initial ideal was to foster sustained collaboration, this was not something that the original researchers necessarily wanted, expected, or could accommodate. Some had left academia for new ventures or were not available. Others had developed different interests and had moved on from their Timescapes work. A few were still using their own project material, and some did not want any contact at all. Whilst original team members may wish to collaborate they may not have the time or funds to do so. Yet, it may well be junior and field researchers who are best placed to enlighten secondary analysts on the minutiae of a project. We were also concerned that sustained collaboration, which relies largely on the goodwill of colleagues, could result in exploitation. It is important to acknowledge the hidden labour involved in such collaborations and to think through the possibilities, where feasible, of formalising the process. This could involve a variety of options from acknowledging the investments of data generators in project outputs through to developing joint ventures or incorporating those willing into applications for funding, especially to recompense for time input. This might be particularly appealing for fixed-term contract researchers.

Case Study 8.2 The Everyday Childhoods Project

Liam Berriman

The Everyday Childhoods project is an example of how data re-use can be anticipated and prospectively embedded within a study from its conception. As a research team with a strong background in qualitative longitudinal research, we were keenly aware of the value of starting with the archive to ensure that future data re-use was negotiated with our participants from the beginning. As a study with children (aged from 7–15 years) and parents/carers as key stakeholders in the research, we wanted to ensure that any plans for data re-use were acceptable to them and that they were fully supported in understanding what the implications of being 'archived' might be. Our study involved several phases of fieldwork that initially lasted for 12 months, during which we regularly had conversations with participants and their families about study's ethical practices, including around consent, anonymity, and data management. At the end of the 12 months, we conducted a follow-on study (Curating Childhoods) which further extended these conversations by funding a workshop at the Keep Archive where the digital data set was due to be archived. These conversations enabled collaborative dialogues around data, in which we introduced participants to research and archiving practices, as well as inviting them to develop their own ethical terminology (which included 'respect' and 'care' for data), as well as writing postcards to future archive users.

Reflecting back on our approach, we have developed a set of four principals for prospective collaborative research that starts with the archive (see Thomson & Berriman, 2021 for a more extended discussion). These principals are not intended to be adopted wholesale, but rather adapted, contested and re-imagined by other studies where archiving data and making it available for re-use is an anticipated outcome.

- Possibility—Research should begin with the question of what the future archive might look like and how possibilities for data re-use can be built into research from the beginning.
- Co-production—Researchers should ensure that they engage in dialogues with stakeholders about the creating an archive available for data re-use from the start of the research process. This should include research participants (which in our case included children and their families) and archivists. Co-production can be a useful way of creating a community around a data archive.
- Shareability—Of working with participants to imagine how and when data will be made publicly available, and who might be the audiences and users of their data.
- Posterity—Participants should be supported in critical and open reflections on what it means to consent to not only participate in a study in the present, but also other potential future studies that may use their data.

The Everyday Childhoods data set is available as an open access archive on Figshare: https://sussex.figshare.com/Everyday_Childhoods and the project's book is available open access via Bloomsbury: https://doi.org/10.5040/9781350011779

Case Study 8.3 Linked Lives and Linked Data: A QLR Study of Journeys through Genomic Medicine

Susie Weller, Kate Lyle, and Anneke Lucassen

'Journeys through genomic medicine' is one of five projects examining the ethical and social challenges that arise for those living and working with genomic results, and forms part of the Ethical Preparedness in Genomic Medicine (EPPiGen) study (Wellcome Trust, 2018–2023; Ref: 208053/B/17/Z). Employing a qualitative longitudinal research design, we have been following the experiences of individuals and families as they undertake testing to determine possible genetic causes or heritable risks for a rare disease or cancer. At the time of writing, the project is ongoing, and we plan to use the breadth-and-depth method to bring this data set into conversation with other existing data sets, as well as making the anonymised interview transcripts available for others to use.

There are two key concerns on which we have reflected as we plan for this endeavour. The first relates to the nature of the work we are conducting. Some participants, for example, speak of rare diseases where there may only be a few cases nationally or even internationally. We are, of course, careful to describe these in more generic terms but by bringing qualitative data sets on the subject together it is possible that those involved in small support groups or those who are connected via social media may recognise the stories of others and, in so doing, identities will be revealed. These concerns highlight the relevance of inadvertent linkages to publicly available data outside of a project.

The second concern relates to the inherent connectedness of 'linked lives' in genomic medicine (Weller et al., 2022). Given that a genetic test result in one person may be relevant to their close relatives, these relatives may have an interest in knowing about the health care of others. Accordingly, multiple members of the same family have often participated, either together or in separate interviews. Maintaining confidentiality within families is important, but also difficult since family members may make connections. We've created multiple pseudonyms for some individuals so that we can link cases within our own analysis but prevent family members discovering confidential information via this route, as well as avoiding such linkages by future users.

References

Weller, S., Lyle, K., & Lucassen, A. (2022). Re-imagining 'the patient': Linked lives and lessons from genomic medicine. *Social Science & Medicine, 297*, 114806. https://doi.org/10.1016/j.socscimed.2022.114806

▶ Case Study 8.4 Shifting Connections to Data in the Timescapes Qualitative Longitudinal Archive

Susie Weller and Rosalind Edwards.

We were involved in the Timescapes initiative, the first large-scale qualitative longitudinal study to be funded in the UK by the Economic and Social Research Council, from the outset. The study comprised seven empirical projects, each focusing on a different lifecourse phase. Our project—'*Siblings and Friends: The changing nature of children's lateral relationships*'—was a 12-year endeavour that followed the lives of 50 young people from mid-childhood into early adulthood (https://timescapes-archive.leeds.ac.uk/timescapes/research/siblings-and-friends/). Whilst acknowledging that our presence in participant's lives may have been experienced by them as intermittent or fleeting, we felt we developed enduring relationships with participants. This sense of connectedness was marked by the moments we shared in their homes, and with their families, and through multiple discussions about their lives over time. We continued to work with the data after the funding ended, whilst also immersed in new projects. Nevertheless, the ebb and flow of a sense of connectedness is still apparent and we have reflected repeatedly about whether this could be explained in terms of a continuing sense of data ownership or whether it was about how our lives had become linked through the moments shared, our developing research relationships, and our repeated engagement with participant's accounts.

We have since developed the 'Working across qualitative longitudinal studies: A feasibility study looking at care and intimacy' project. The study presented us with the chance to re-visit the 'Siblings and Friends' data as well as engage with data from five of the other Timescapes studies. As part of our focus on developing secondary analytic practice, we consulted colleagues about their feelings towards our planned reuse of the data sets. Although some were still using the material, others felt their connection to the data had diminished over time. For us, having had the opportunity to work with the data for four years meant that we felt a growing sense of connection to participants and their narratives, as well as, to the endeavours of the original researchers. We became attached to it as our production, which altered our perceptions of ownership (Weller, 2022). We created and deposited a new data set in the Timescapes Data Archive, which comprised our assemblage of qualitative data from the six studies (Davidson et al., 2019). In these terms, (re)making research data available for others to use suggests that knowledge production is open-ended and muddies the waters of data ownership (Weller & Edwards, 2022).

References

Davidson, E., Edwards, R., Jamieson, L., & Weller. (2019). Big data, qualitative style: A breadth-and-depth method for working with large amounts of secondary qualitative data. *Quality & Quantity, 53*(1), 363–376.

Weller, S. (2022). Fostering habits of care: Reframing qualitative data sharing policies and practices. *Qualitative Research*. https://doi.org/10.1177/14687941211061054

Weller, S., & Edwards, R. (2022). Ownership, connectedness and archives: Changing perceptions over time. *Timescapes Blog Series*. https://timescapes-archive.leeds.ac.uk/ownership-connectedness-and-archives-changing-perceptions-over-time/

References

Ballard, L. (2020). *Wellbeing of psychologists during Covid-19: impact on research* https://www.youtube.com/watch?v=u8DGEW16GZ4

Carroll, S. R., Garba, I., Figueroa-Rodríguez, O. L., Holbrook, J., Lovett, R., Materechera, S., Parsons, M., Raseroka, K., Rodriguez-Lonebear, D., Rowe, R., Sara, R., Walker, J. D., Anderson, J., & Hudson, M. (2020). The CARE principles for indigenous data governance. *Data Science Journal, 19*(1), 43. https://doi.org/10.5334/dsj-2020-043

Chilisa, B., Major, E. T., & Khudu-Petersen, K. (2017). Community engagement with a postcolonial, African-based relational paradigm. *Qualitative Research, 17*(3), 326–339.
Corti, L. (2007). Re-using archived qualitative data—Where, how, why? *Archival Science, 7*, 37–54.
Davidson, E., Edwards, R., Jamieson, L., & Weller, S. (2019). Big data, qualitative style: A breadth-and-depth method for working with large amounts of secondary qualitative data. *Quality & Quantity, 53*(1), 363–376.
Gillies, V., & Edwards, R. (2012). Working with archived classic family and community studies: Illuminating past and present conventions around acceptable research practice. *International Journal of Social Research Methodology, 15*(4), 321–330. https://doi.org/10.1080/13645579.2012.688323
Glenna, L., Hesse, A., Hinrichs, C., Chiles, R., & Sachs, C. (2019). Qualitative research ethics in the big data era. *The American Behavioral Scientist, 63*(5), 560–583.
Hardy-Johnson, P., Weller, S., Kehoe, S. H., Barker, M., Haileamalak, A., Jarju, L., Jesson, J., Krishnaveni, G. V., Kumaran, K., Leroy, V., Moore, S., Norris, S. A., Patil, S., Sahariah, S., Ward, K., Yajnik, C., Fall, C., & the TALENT collaboration. (2021, November). Exploring adolescent diet and physical activity in India and sub-Saharan Africa. *Field Exchange, 66*, 61. www.ennonline.net/fex/66/adolescentdietphysicalactivity
Harford, T. (2014). Big data: A big mistake? *Significance, 11*, 14–19. https://doi.org/10.1111/j.1740-9713.2014.00778.x
Hughes, K., & Tarrant, A. (2020). The ethics of qualitative secondary analysis. In K. Hughes & A. Tarrant (Eds.), *Qualitative secondary analysis*. Sage.
Keikelame, M. J., & Swartz, L. (2019). Decolonising research methodologies: Lessons from a qualitative research project, Cape Town, South Africa. *Global Health Action, 12*(1), 1561175. https://doi.org/10.1080/16549716.2018.1561175
Lucivero, F. (2020). Big data, big waste? A reflection on the environmental sustainability of big data initiatives. *Science and Engineering Ethics, 26*, 1009–1030. https://doi.org/10.1007/s11948-019-00171-7
Lyle, K., Horton, R., Weller, S., & Lucassen, A. (2023). Immortal data: A qualitative exploration of patients' understandings of genomic data. *European Journal of Human Genetics*. 31, 681–686.
Lyle, K. Weller, S., Samuel, G., & Lucassen, A. (2022, June 20). Beyond regulatory approaches to ethics: Making space for ethical preparedness in healthcare research. *Journal of Medical Ethics*. Published Online First. https://doi.org/10.1136/medethics-2021-108102.
Mason, J. (2007). "Re-using" qualitative data: On the merits of an investigative epistemology. *Sociological Research Online, 12*(3), 1–4. http://www.socresonline.org.uk/12/3/3.html
Mauthner, N. (2012). Are research data a common resource? *Feminists@Law, 2*(2), 1–22.
Mills, K. A. (2018). What are the threats and potentials of big data for qualitative research? *Qualitative Research, 18*(6), 591–603. https://doi.org/10.1177/1468794117743465
Mittelstadt, B. D., Allo, P., Taddeo, M., Wachter, S., & Floridi, L. (2016). The ethics of algorithms: Mapping the debate. *Big Data & Society, 3*(2), 10.1177/2053951716679679.
Moore, N., Salter, A., Stanley, L., & Tamboukou, M. (2017). *The archive project: Doing archival research in the social sciences*. Routledge.
Neale, B. (2021). *The craft of qualitative longitudinal research*. Sage.
Nind, M., Meckin, R., & Coverdale, A. (2021). *Changing research practices: Undertaking social research in the context of Covid-19: Project Report* NCRM. https://eprints.ncrm.ac.uk/id/eprint/4457/
Pascoe Leahy, C. (2021). The afterlife of interviews: Explicit ethics and subtle ethics in sensitive or distressing qualitative research. *Qualitative Research, 22*(5), 777–794. https://doi.org/10.1177/14687941211012924
Research Data Alliance International Indigenous Data Sovereignty Interest Group. (2019, September). *CARE principles for indigenous data governance*. The Global Indigenous Data Alliance. GIDA-global.org. gida-global.org/care
Roulston, K. (2017). *Research integrity and the qualitative researcher*. Retrieved January 24, 2022, from https://qualpage.com/2017/03/09/research-integrity-and-the-qualitative-researcher/

Samuel, G., & Lucivero, F. (2020). Responsible open science: Moving towards an ethics of environmental sustainability. *Publications, 8*(4), 54. https://doi.org/10.3390/publications8040054

Sherren, K., Parkins, J. R., Smit, M., Holmlund, M., & Chen, Y. (2017). Digital archives, big data and image-based culturomics for social impact assessment: Opportunities and challenges. *Environmental Impact Assessment Review, 67*, 23–30.

Thambinathan, V., & Kinsella, E. A. (2021). Decolonising methodologies in qualitative research: Creating spaces for transformative praxis. *International Journal of Qualitative Methods, 20*, 1–9.

Thomson, R., & Berriman, L. (2021, June 18). Starting with the archive: Principles for prospective collaborative research. *Qualitative Research*. Epub ahead of print. https://doi.org/10.1177/14687941211023037.

Tisdale, K. (2004). Being vulnerable and being ethical with/in research. In K. B. DeMarrais & S. D. Lapan (Eds.), *Foundations of research: Methods of inquiry in education and the social sciences* (pp. 13–30). Lawrence Erlbaum.

Tizard, B., & Hughes, M. (1984). *Young children learning*. Fontana.

Tuhiwai Smith, L. (2021). *Decolonizing methodologies: Research and indigenous peoples* (3rd ed.). Zed Books.

Walkerdine, V., & Lucey, H. (1989). *Democracy in the kitchen: Regulating mothers and socialising daughters*. Virago.

Walter, M., & Andersen, C. (2016). *Indigenous statistics: A quantitative research methodology*. Routledge.

Watts, J. H. (2008). Integrity in qualitative research. In L. M. Given (Ed.), *The sage encyclopaedia of qualitative research methods* (Vol. 1, pp. 440–441). Sage Publications.

Weller, S. (2023). Fostering habits of care: Reframing qualitative data sharing policies and practices. *Qualitative Research, 23*(4), 1022–1041.

Weller, S., & Edwards, R. (2022). Ownership, connectedness and archives: Changing perceptions over time. *Timescapes Blog Series*. https://timescapes-archive.leeds.ac.uk/ownership-connectedness-and-archives-changing-perceptions-over-time/

Weller, S., Lyle, K., & Lucassen, A. (2022, February 12). Re-imagining 'the patient': Linked lives and lessons from genomic medicine. *Social Science & Medicine, 297*, 114806. https://doi.org/10.1016/j.socscimed.2022.114806.

Zarate, O. A., Brody, J. G., Brown, P., Ramirez-Andreotta, M. D., Perovich, L., & Matz, J. (2016). Balancing benefits and risks of immortal data: Participants' views of open consent in the personal genome project. *The Hastings Center Report, 46*(1), 36–45. https://doi.org/10.1002/hast.523

Big Qual and the Future of Qualitative Analysis

9

9.1 Introduction

In this final chapter, we conclude by identifying key moments in the development of big qual, and the place of the breadth-and-depth method within this. We pause to reflect on:

- the emergence of big qual and the associated epistemological and methodological values
- the breadth-and-depth method as a process
- the opportunities, challenges, and limitations this way of working can bring
- the implications of the method, both for the field of qualitative enquiry and for big data debates

9.2 The Emergence of Big Qual

The breadth-and-depth method was developed at an important moment in the field of qualitative enquiry. In part, it was a response to the growing availability of qualitative material for reuse in international contexts. This was being driven by moves to promote open access science and supported by policy initiatives such as the *OECD Principles and Guidelines for Access to Research Data from Public Funding* (OECD, 2007) and the *Berlin Declaration on Open Access to Knowledge in the Sciences and Humanities* (Berlin Declaration, 2003). Integral was a concern that investments in publicly funded research ought to be maximised, with greater emphasis being placed on data auditing and accountability (Carusi & Jirotka, 2009; Slavnic, 2017; Weller, 2022). As a result, in contexts such as the UK, research institutions, publishers, and funding bodies have increasingly required researchers to make data available for reuse. This shift has been significant for qualitative data where archiving and reuse remained fairly infrequent.

© The Author(s), under exclusive license to Springer Nature Switzerland AG 2023

S. Weller et al., *Big Qual*, https://doi.org/10.1007/978-3-031-36324-5_9

At the same time there has been growth in the digital infrastructures and technical capabilities to support the curation, retention, and sharing of such material (Glenna et al., 2019; Van den Eynden & Corti, 2020). Formal qualitative archives associated with university libraries and research institutions are gaining ground both in terms of size and esteem. Meanwhile, secondary qualitative data reuse has evolved into a diverse range of methodological approaches (Hughes & Tarrant, 2020). There are also a growing number of smaller community-based archives that are making space for social groups to document, record, and take control of their own knowledge archives (Moore et al., 2017).

Through these advances, qualitative secondary analysis has gained ground. However, big qual, both as a concept and practice, has remained a relatively uncharted terrain. The influence of big data was initially not of interest to qualitative researchers. Rather, it was seen as the domain of those working in the field of computational science, with its value being attributed to 'bigness'. Concerns, however, were increasingly voiced about the ability of computational tools to make sense of data on social life and social processes. There is now a growing recognition that big data does not only equate to numbers, but also an array of new forms of qualitative data, ranging from social media to online forums and other digital communications. Qualitative enquiry is gradually being recognised as having a critical role within big data, with recognition that its philosophy and analytical tools are necessary for facilitating understanding about social context, processes, subjectivities, and complexity. At the same time, by bringing qualitative enquiry into big data analytics there is an opportunity to address growing concerns that biases and structural inequalities are being reproduced (Benjamin, 2019) and positivist epistemologies privileged (Glenna et al., 2019).

Bringing qualitative forms of enquiry into the analysis of big data has generated new possibilities in terms of the insight gained into big data and the opportunities borne from dialogue and collaboration between researchers who consider themselves qualitative and those who view themselves as computational or quantitative. Those at the vanguard of these approaches (Mills, 2019; Wiedemann, 2016) are challenging unhelpful borderlines between qualitative, quantitative, and computational research. This has given space for projects that epistemologically work across "methodological borderland[s]" (Garcia & Ramirez, 2021: 240), and value integrative, collaborative, and multi-disciplinary ways of working.

The breadth-and-depth method sits within, but extends, this body of work. It is not, as we have described, an approach for working qualitatively with big data, although it has grown from this body of work. Rather, this is a methodologically integrative big qual strategy for enabling researchers to gather, handle, combine, and analyse large volumes of qualitative data. By raising the profile of the range of qualitative data sources available to researchers, the method is challenging assumptions about what constitutes 'big' data. Big data is not only digitally borne, nor does it necessarily consist of many terabytes of data. Instead, the method presented in this book encourages exploration of the wealth and diversity of 'big' qualitative data sources and seeks to illuminate the possibilities that come from combining big data analytics with qualitative ways of thinking.

9.3 Overview of the Breadth-and-Depth Method

The breadth-and-depth method is the result of a four-year endeavour—"Working across qualitative longitudinal studies: A feasibility study looking at care and intimacy" funded by the UK's National Centre for Research Methods. The study examined the possibilities for developing new procedures and extending good practice for working across multiple sets of archived qualitative data. In short, we wanted to know whether it was possible to do big qual analysis while retaining all that is distinct about rigorous qualitative research.

Our task was to develop a means of reconciling breadth with depth. It is simply impractical to conduct big qual manually, analysing in-depth thousands of detailed text files. Instead, we needed to survey the data landscape to get a sense of the available material and to identify, in a rigorous manner, appropriate places to dig deeper. We realised that the process was much like that undertaken by a field archaeologist, with a collection of big qual data sets constituting a landscape that must be scanned to gain a sense of the breadth of the material and to pinpoint features of potential interest. By adopting an archaeological metaphor, we were able to consider how we access data, at different levels and in alternative ways, and to think about what lies beneath the corpus of material being analysed. It helped us to work both extensively and intensively to identify and excavate meaning.

The resultant method comprises a four-step process fusing computational techniques for exploring breadth with more conventional forms of qualitative analysis to provide depth. Step one—*aerial surveying*—commences with processes akin to aerial reconnaissance with the researcher flying systematically across a data landscape. A map (i.e. set of projects) aids the researcher in gaining an overview of the textures and features of different land masses (i.e. projects). In so doing, you could bring together data from many different projects drawing on material that has been digitised and housed in a research repository or material pooled as part of a collaboration with other researchers. The aim is to locate and review potential sources of academic data of an appropriate nature, quality, and 'fit' with the research topic. These selected data become the composite data set with which you work.

Once the data is surveyed, step two comprises a ground-based *geophysical survey* of an area (i.e. collection of data sets) to assess what merits closer investigation. This step involves conducting breadth analysis using computational text mining tools to identify the key concepts, associated with the researchers' interests, across the composite data set. The aim of this step is to locate areas of conceptual and substantive interest. Particular sub-surface features are examined to determine the most important places to dig deeper into the data. Importantly, the outcomes of this recursive thematic mapping are not the end point. Rather, just as an archaeologist would view a geophysical survey as a means to inform where to create test pits, so step two is concerned with using computational tools to determine where to dig deeper into the data.

The results produced by data mining are the starting points for sampling short extracts of data in step three, which is concerned with conducting preliminary analysis or *shallow test pit sampling*. Test pits are created where it is thought, based on the previous surveys, artefacts of interest (i.e. data of relevance to a research question/project) may be found. Using the outcomes from step two, short extracts from the data—or test pits—are explored to see which areas of interest are worth investigating further. The sampling logic employed is shaped by the research questions and design. If the test pit does not yield any material of salience to your research, then you can return to step two to locate alternative areas of interest. Even at this shallow stage, it is important to keep the context in which extracts were generated in mind. Together, steps two and three are a systematic, rigorous, and considered way of moving from breadth to depth.

The final task—step four—concerns archaeological *deep excavation* exploring in detail specific cases of interest, focusing on depth rather than breadth. This step involves moving from examining extracts of data to immersion in whole cases. You can employ one of a number of commonly used approaches to qualitative analysis including thematic, frame, discourse, or narrative analysis. It involves being sensitive to context and complexity. This step may reveal other issues for exploration too—so again you could go back to step two.

9.4 Stepping out of the Methodological Borderlines

As the preceding chapters have demonstrated, the breadth-and-depth method is an iterative process, with each step guided by the outcomes of the one previous. At any point, you can go back and re-survey the landscape to enhance your understanding of the social processes at work. Analysis is conducted iteratively through a simultaneous zooming out to the aerial view and zooming in to the in-depth view. Through an integration of computational text analysis and in-depth qualitative analysis, both approaches are brought into an active dialogue resulting in both extensive coverage and intensive illumination (Davidson et al., 2019, p. 368).

In our reflections on the breadth-and-depth method, and from the numerous case studies presented, it is clear that the method offers ways of seeing not previously possible. As social researchers, we still too often seek to explain social phenomena from our own disciplinary perspective. This can be the result of our initial training, the desire to specialise across our career trajectories, or a consequence of institutional bureaucracies which limit inter-disciplinary working. Even those committed to mixed methods can find themselves facing limitations, with qualitative research too often devalued as a mere 'supplement' to quantitative forms of data. At the other end of the spectrum, quantitative researchers may wish to engage in qualitative forms of enquiry, but feel unable due to a lack of skills, training or temperament. We recognise and value the need for different approaches to research—working at different scales is a necessity. However, we also see enormous value in integrative approaches that encourage and support researchers to step outside of their established methodological borderlines. The breadth-and-depth method provides a unique opportunity to do this.

One of the central gains from using the breadth-and-depth method is its capacity to bring comparison into the research design by combining multiple qualitative data sets. This allows for greater diversity of participants and cases, generating the opportunity to work at a level above that typically possible from a single study. We have argued that working in this way offers a theoretical approach to generalisation. This is not a generalisability based on gaining a representative or statistically significant sample. Rather, the method affords a generalisability that comes from particularity. By increasing both the number of participants and the diversity of the sample there is potential for researchers to gain insight into the complexity of social processes at different scales and in different contexts, and in how they relate, and unfold (Davidson et al., 2019). Through diversity within a sample of multiple sets of data, there also comes the possibility of softening the complaint that qualitative forms of enquiry make 'unwarranted assumptions [...] about the characteristics of the population of cases not yet studied' (Seale, 1999, p. 112).

By promoting the breadth-and-depth method, we do not wish to dilute or devalue the status of qualitative research, or replace its philosophy or values. Rather, the unique aspect of the breadth-and-depth method is its capacity to bring together different theoretical, disciplinary, and epistemological perspectives. Such integration can come from the different approaches that shape the projects and data brought together. Combining projects from different locations, contexts, and timeframes can generate innovative insights. Throughout the book, we have sought to ensure that we do not present the breadth-and-depth method as offering better understandings of the social world. Rather, our claim is that it generates new and different types of understandings that have the capacity to travel across both macro and more granular scales.

Integration, of course, also comes from the unique way in which the breadth-and-depth method enables social researchers across the qualitative–quantitative spectrum to gain new insights from the accumulating large volumes of qualitative data. To do this, the method combines computational techniques and analytical approaches typically (and erroneously) labelled 'quantitative' with approaches that are unequivocally 'qualitative'. Researchers in both traditions must, therefore, leave their comfort zones and, in the process, learn new skills and perhaps also new ways of thinking. While there is some distance to go in terms of narrowing methodological approaches, the breadth-and-depth method holds the potential for cultivating greater cross-over in techniques and constructive collaboration and dialogue.

9.5 Challenges and Limitations

While we are keen to emphasise the possibilities and opportunities for extending big qual methodologies, we are also acutely aware of its challenges and limitations. The first issue is one of pragmatics—there may be a substantive topic that would benefit from this type of investigation, but the data may simply not be 'out there' (Mauthner et al., 1998). Alternatively, the data may exist, but not be available or accessible. These issues will partly be addressed as the archive project—internationally, nationally, and locally—continues to grow and expand.

That being said, we also wish to raise caution against the open science movement. Moves to archive and make data available "by default" (Thomson & Berriman,

2021, p. 11) raise a number of issues for qualitative researchers seeking to bring together large volumes of qualitative material from multiple sources, as well as quantitative researchers involved in, for example, survey work. Data sharing agreements are important, but the danger remains that, if not practised with care, the contributions of researchers and/or participants could, albeit unintentionally, be exploited. While such a caution applies to all forms of data collection and analysis, arguably the risk is heightened when multiple data sets, potentially from different sources and with different agreements, are brought together.

Issues of data sovereignty speak more widely to concerns about the level of control participants have over their data over time. Since this is a new method, participants may not have foreseen the use of their data in this way. So, while the breadth-and-depth method has the potential to contribute socially impactful research, care must be taken in ensuring that re-use encompasses practices of respect, care, and collectivity (Carroll et al., 2020). Decolonising and Indigenous approaches to research are useful reflective tools here, helping to raise sets of questions around ownership, control, power, and benefits.

There is also a need to be cognisant of the possibility that those using the breadth-and-depth method still give epistemic privilege to big data analytics. This might mean that greater regard or confidence is attributed to computational outputs. While we encourage researchers to adapt the method according to their own study and epistemology, such analytical practices would be in opposition to the integrative ethos that underpins the approach set out in the book. This ethos is one that celebrates collaboration, integration, and interdisciplinary dialogue.

9.6 Moving the Integrative Field Forward

While our aim is to move the integrative field forward, we recognise that our support for computational textual techniques will be a hard ask for some qualitative researchers. At the same time, our instruction to return to more conventional qualitative methods may be disappointing to those seeking a quick way of analysing more qualitative data than can easily be read by a human. It may also be a hard ask for computational social scientists and quantitative researchers working with big data to bridge into a more inclusive approach to identifying and analysing data.

Our position is that oppositions and methodological dualisms are not only unhelpful (Rieder & Röhle, 2017), but they are ahistorical since they fail to recognise long-standing alternative positions. There is—without question—as much 'craft' associated with quantitative work as there is qualitative. Similarly, qualitative research is equally able to come from a positivist tradition as quantitative research. We firmly believe that in learning each other's crafts, both stand to gain. This is being demonstrated through big qual, a field which has developed to include a diverse range of combinations crossing between working qualitatively and quantitatively—quali-quan methods, as well as innovative methods for collecting and analysing large volumes of qualitative analysis. These approaches are generating integrative spaces, where methodological bridges can be built, not re-ascribed (boyd & Crawford, 2011).

Undoubtedly, big data raises longstanding epistemological questions, and these require serious, critical attention. The emergent field of big qual is bringing these issues to light. The challenge for qualitative research is in finding ways to engage meaningfully with big data and broaden the possibilities it might offer. We hope that in time that qualitative forms of enquiry are routinely acknowledged in big data analytics. In our own search for breadth and depth, and in developing ways of analysing large volumes of qualitative data, our imagining was for a new approach which, rather than re-inscribe methodological divisions, might bring big data and qualitative philosophies and approaches together. We have set out our case that investment and institutional support should be given to training, capacity building, and multi-disciplinary working, and that this should happen across the methodological spectrum. In other words, this should involve computational scientists and quantitative researchers gaining experience and expertise in qualitative approaches and sensibilities, and qualitative researchers developing skills in computational techniques.

This book has been a demonstration of a method in practice. Through our own project we have documented and illustrated a way of analysing and representing large volumes of qualitative data, and doing so with integrity and multi-disciplinary values. By providing a range of case studies and resources, we also hope to situate the method in diverse cultural, social, and geographical settings. While our own project is concerned with sociological understandings of intimate relations, care, and gender, we have also sought to emphasise that the approach can be applied across substantive fields, and from different methodological disciplines. The creation of the breadth-and-depth method establishes a significant point in the history of qualitative enquiry. Having a practical method that enables researchers to gather, handle, combine, and analyse large volumes of qualitative data also has enormous potential in policy domains. It does this by ensuring that qualitative data, and most critically qualitative forms of enquiry, are not lost, neglected, or ignored by big data analytics. Rather, it foregrounds the necessity of having analytical processes that can discern breadth through high-level patterns, trends, and associations, alongside nuanced, contextualised, in-depth understandings. The development of big qual to date demonstrates that there is a huge appetite amongst researchers, not only for the method, but more generally for new and innovative ways of exploring and reusing the range of qualitative data available. We hope that the breadth-and-depth method is only the beginning of this journey.

9.7 Resources

The following papers and book chapters provide further details regarding the breadth-and-depth method:

Davidson, E., Edwards, R. Jamieson, L., & Weller, S. (2019). Big data, qualitative style: A breadth-and-depth method for working with large amounts of secondary qualitative data. *Quality & Quantity*, *53*(1), 363–376.

Edwards, R., Davidson, E., Jamieson, L., & Weller, S. (2021). Theory and the breadth-and-depth method of analysing large amounts of qualitative data: A research note. *Quality & Quantity, 55*, 1275–1280.

Edwards, R., Weller, S., Jamieson, L., & Davidson, E., (2020). Search strategies: Analytic searching across multiple data sets and with combined sources, In K. Hughes, & A. Tarrant (Eds.), *Qualitative secondary analysis* (pp. 79–100). Sage.

Weller, S., Davidson, E., Edwards, R., & Jamieson, L. (2019). Analysing large volumes of complex qualitative data: Reflections from international experts. N*CRM Working Paper*. http://eprints.ncrm.ac.uk/4266/

The following podcasts provide an overview of the method:

Digging deep! The archaeological metaphor helping researchers get into big qual. https://www.ncrm.ac.uk/resources/podcasts/mp3/NCRM_podcast_Weller2.mp3

Making space for Big Qual: New ideas in research methods and teaching. https://www.ncrm.ac.uk/resources/podcasts/?id=space-for-big-qual

These resources have been designed to support teaching of the breadth-and-depth method:

Weller, S., Davidson, E., Edwards, R., & Jamieson, L., (2019). *Big qual analysis: Teaching data set*. Timescapes Archive. https://doi.org/10.23635/14

Lewthwaite, S., Jamieson, L., Weller, S., Edwards, R., & Nind, M. (2019). Teaching how to analyse large volumes of secondary qualitative data. *NCRM Online Learning Resource*. https://www.ncrm.ac.uk/resources/online/all/?id=20727

The following website contains a wide range of resources relating to big qual analysis:

Big Qual Analysis Resource Hub: www.bigqlr.ncrm.ac.uk

References

Berlin Declaration. (2003). *Berlin declaration on open access to knowledge in the sciences and humanities*. https://openaccess.mpg.de/Berlin-Declaration

Benjamin, R. (2019). *Race after technology: Abolitionist tools for the new jim code*. Cambridge: Polity Press.

boyd, d., & Crawford, K. (2011). Six provocations for big data. A decade in internet time: Symposium on the dynamics of the internet and society. https://doi.org/10.2139/ssrn.1926431.

Carroll, S. R., Garba, I., Figueroa-Rodríguez, O. L., Holbrook, J., Lovett, R., Materrechera, S., et al. (2020). 'The CARE Principles for Indigenous Data Governance'. *Data Sci. J., 19*(43), 1–12.

Carusi, A., & Jirotka, M. (2009). From data archive to ethical labyrinth. *Qualitative Research, 9*(3), 285–298.

Davidson, E., Edwards, R., Jamieson, L., & Weller, S. (2019). Big data, qualitative style: A breadth-and-depth method for working with large amounts of secondary qualitative data. *Quality & Quantity, 53*(1), 363–376. Open access.

Garcia, G. A., & Ramirez, J. J. (2021). Proposing a methodological borderland: Combining chicana feminist theory with transformative mixed methods research. *Journal of Mixed Methods Research, 15*, 240–260.

Glenna, L., Hesse, A., Hinrichs, C., Chiles, R., & Sachs, C. (2019). Qualitative research ethics in the big data era. *The American Behavioral Scientist, 63*(5), 560–583. https://doi.org/10.1177/0002764218805806

Hughes, K., & Tarrant, A. (Eds.). (2020). *Qualitative secondary analysis*. Sage.

Mauthner, N., Parry, O., & Milburn, K. (1998). The data are out there, or are they? Implications for archiving and revisiting qualitative data. *Sociology, 32*, 733–745.

Mills, K. A. (2019). Big data for qualitative research (1st ed.). Routledge.

Moore, N., Salter, A., Stanley, L., & Tamboukou, M. (2017). *The archive project: Doing archival research in the social sciences*. Routledge.

OECD. (2007). *OECD principles and guidelines for access to research data from public funding*. https://www.oecd.org/sti/inno/38500813.pdf

Rieder, B., & Röhle, T. (2017). Digital methods: From challenges to Bildung. In M. Schäfer & K. Van ES. (Eds.), *The datafied society: Studying culture through data*. Amsterdam University Press.

Seale, C. (1999). *The quality of qualitative research*. Sage.

Slavnic, Z. (2017). Research and data-sharing policy in Sweden—Neoliberal courses, forces and discourses. *Prometheus, 35*(4), 249–266.

Thomson, R., & Berriman, L. (2021, June 18). Starting with the archive: Principles for prospective collaborative research. *Qualitative Research*. Epub ahead of print. https://doi.org/10.1177/14687941211023037.

Van den Eynden, V., & Corti, L. (2020). The importance of managing and sharing research data. In L. Corti, V. Van den Eynden, L. Bishop, & M. Woollard (Eds.), *Managing and sharing research data: A guide to good practice* (pp. 1–32). Sage.

Weller, S. (2022). Fostering habits of care: Reframing qualitative data sharing policies and practices. *Qualitative Research*. https://doi.org/10.1177/14687941211061054.

Wiedemann, G. (2016). *Text mining for qualitative data analysis in the social sciences*. Springer Fachmedien Wiesbaden.

Author Index

© The Author(s), under exclusive license to Springer Nature Switzerland AG 2023

S. Weller et al., *Big Qual*, https://doi.org/10.1007/978-3-031-36324-5

Subject Index

© The Author(s), under exclusive license to Springer Nature Switzerland AG 2023

S. Weller et al., *Big Qual*, https://doi.org/10.1007/978-3-031-36324-5